CW01672860

Handbook of Clinical Techniques in the Management of Chronic Pain

Handbook of Clinical Techniques in the Management of Chronic Pain

J.R. Wedley MB Ch.B FRCA DA
Consultant Anaesthetist
Pain Relief Clinic, Guy's Hospital, London, UK

and

C.A. Gauci MD FRCA
Consultant Anaesthetist
Pain Relief Clinic, Whipps Cross Hospital
London, UK

harwood academic publishers
Australia • Canada • China • France • Germany • India • Japan • Luxembourg • Malaysia
The Netherlands • Russia • Singapore • Switzerland • Thailand • United Kingdom

Copyright © 1994 by Harwood Academic Publishers GmbH.

All rights reserved.

First published 1994
Second printing 1996

No part of this book may be reproduced or utilized in any form or by any means, electronic or mechanical, including photocopying and recording, or by any information storage or retrieval system, without permission in writing from the publisher. Printed in Singapore.

Emmaplein 5
1075 AW Amsterdam
The Netherlands

While the information in this book is believed to be true and accurate at the date of going to press, neither the author nor the publisher can accept any legal responsibility for any injury and/or damage to persons or property as a matter of products liability, negligence or otherwise or form any use or operation of any methods, products, instructions or ideas contained in the material herein. Because of rapid advances in the medical sciences, the publisher recommends that independent verification of diagnoses and drug doses should be made.

British Library Cataloguing in Publication Data

A catalogue record for this book is available from the British Library.

ISBN 3-7186-5399-0 hardcover
90-5702-067-X softcover

Contents

To

Colonel J. McEwan OSJ FRCA L/RAMC

who taught us both so much
during the early years of our careers
in the field of pain relief therapy
we dedicate this book.

Preface

Chronic pain relief therapy is now an important speciality in its own right; pain clinics are appearing in many hospitals, both in the UK and abroad, and more doctors, usually but not invariably anaesthetists, are being appointed to run these clinics and work in them. Very often doctors taking up consultant posts 'with an interest in pain' have little practical experience of the many procedures now available to treat their patients.

We have been working in the field of chronic pain relief therapy for several years and have instructed a number of doctors in various pain relief techniques. Both of us have been struck by the absence of any books which describe in clear and simple steps the most commonly performed pain relief procedures in use at the present time. There are, of course, many excellent books on local anaesthetic techniques as well as many textbooks on the various theoretical aspects of pain relief, and several of the latter describe, in general terms, a number of procedures used for the treatment of chronic pain. We feel, however, that there is a need for a relatively inexpensive book written primarily for the 'pain relief doctor' working on his own in a district general hospital: a book which concentrates exclusively on the practical side of the speciality and which assumes that the reader is inexperienced and has to be guided step by step through the various procedures until he acquires enough experience to dispense with it altogether.

In *A Handbook of Clinical Techniques in the Management of Chronic Pain* we think we have produced such a book. We emphasize that it is based on our own experience and our own thoughts; it is not a distillate of what other workers in the field of chronic pain relief may think or do. Every technique described has been carried out by one or both of us several times.

In an effort to make the book as inexpensive and therefore as widely accessible as possible, we have purposely not covered ground which is well covered in other books. Thus, peripheral nerve blocks and acupuncture are excluded, since there are so many good practical texts available. We consider the intra-articular injection of large joints (shoulder, hip and knee) to lie within the province of the rheumatologist and the specialist in physical medicine, and we have therefore omitted these techniques.

The *Handbook* is intentionally written in a didactic manner and is furnished with many illustrations. It contains very little theory, being essentially a practical guide of the 'how to do it' variety. It is meant to be taken into the operating theatre and left open for ready reference. We hope that this is made easier by the large page size, the binding and the laminated page surfaces. If the reader follows the step-by-step instructions and refers to the illustrations, he will soon become sufficiently experienced to use the book as merely an occasional reference work!

We have assumed that the reader is used to working with an image intensifier; as far as possible we have tried to relate the illustrations to what is actually seen on the X-ray monitoring screen. This should give a clear idea of the various landmarks and steps in each procedure. We have also assumed that the reader is sufficiently familiar with radiation safety. We cannot emphasise too strongly the potential health hazard of X-ray equipment both to the operator and to the patient. Having said this, however, we must point out that several possible complications of 'needling' techniques are preventable by frequent X-ray screening to ascertain at all times the exact position of the needle tip.

All procedures must be carried out using a strict aseptic technique; in some cases we consider antibiotic cover essential. These situations are clearly stated in the text.

Within our self-imposed terms of reference, we feel that we have covered all the commonly used pain relief procedures and we hope that the reader will find our book to be of value in his or her own clinical practice.

J.R. Wedley

C.A. Gauci

1

Radiofrequency

1 General Introduction

Radiofrequency is a widely used method for lesioning nervous tissue, and in the section that follows you will find several commonly used techniques described in some detail. Here, however, we present a brief general outline of the main features of radiofrequency.

The principal temperature-monitored radiofrequency lesion maker commercially available is the RADIONICS equipment (Figure 1). With this technique, a very-high-frequency current is passed down a thermocouple probe (Figure 2) which is inserted through a special cannula, fully insulated except for its tip (Figures 3, 4). When the current is passed down the thermocouple probe, it heats the surrounding tissues to a temperature which can be totally controlled by the operator.

The main advantage of radiofrequency is the total control enjoyed by the operator; thus, the nerve can first be located by passing a stimulating current down the thermocouple probe and then a well-circumbscribed lesion can be induced. Since myelinated fibres are more resistant to heat than are non-myelinated fibres, differential effects can be produced.

The cannula, usually a Sluijter–Mehta radiofrequency cannula, is placed in close proximity to the nerve in question, using X-ray control. The stylet is removed from the cannula and replaced by the thermocouple probe. The operator then attempts to seek the nerve by low-voltage stimulation at a frequency of 50–100 Hz; he seeks the strongest possible sensory stimulation at the lowest possible voltage.

Ford and co-workers (1984) have shown that if the Sluijter–Mehta radiofrequency cannula is actually resting on the nerve, the minimum stimulation required to produce a discharge is 0.5 mA or approximately 0.25 V for a standard 500 Ω tissue impedance (1). As one moves away from the nerve, at a distance of 1 cm 2 V would be required. The cannula needs to be within 3 mm of the nerve in order to create an adequate lesion, and this would be indicated by a maximum stimulation level of around 0.6 V.

The operator should always ensure that his cannula/thermocouple probe is not dangerously close to any motor nerve in the vicinity of the sensory nerve he is trying to lesion. The way this is done, once the sensory nerve has been located, is by stimulating at a low frequency (2 Hz). If no muscle twitch is noted at below 2 V on this frequency, it can be assumed that there are no motor paths within 3 mm of the needle; even if there were, the protective myelin sheath would make them resistant to damage. A safety 'rule of thumb' is that so long as muscle twitches do not appear at less than twice the voltage strength necessary to achieve sensory stimulation, there is no risk of damage to the motor nerve.

When the operator is satisfied that the needle is safely in position, a radiofrequency current (about 300 kHz) is passed through the thermocouple probe. The current heats up the surrounding tissues and produces a lesion in the nerve which has been targeted.

About the Lesion

A current flows from the active tip of the cannula and heat is generated electromechanically by the movement of charged ions in the tissue electrolytes. These ions are pulled to and fro by the current alternating at 300 kHz (i.e. 300 000 times a second) and heat is produced by electromechanical friction. As the current is applied, tissue heating will occur and the lesion will grow until a steady state is reached; at this point the passage

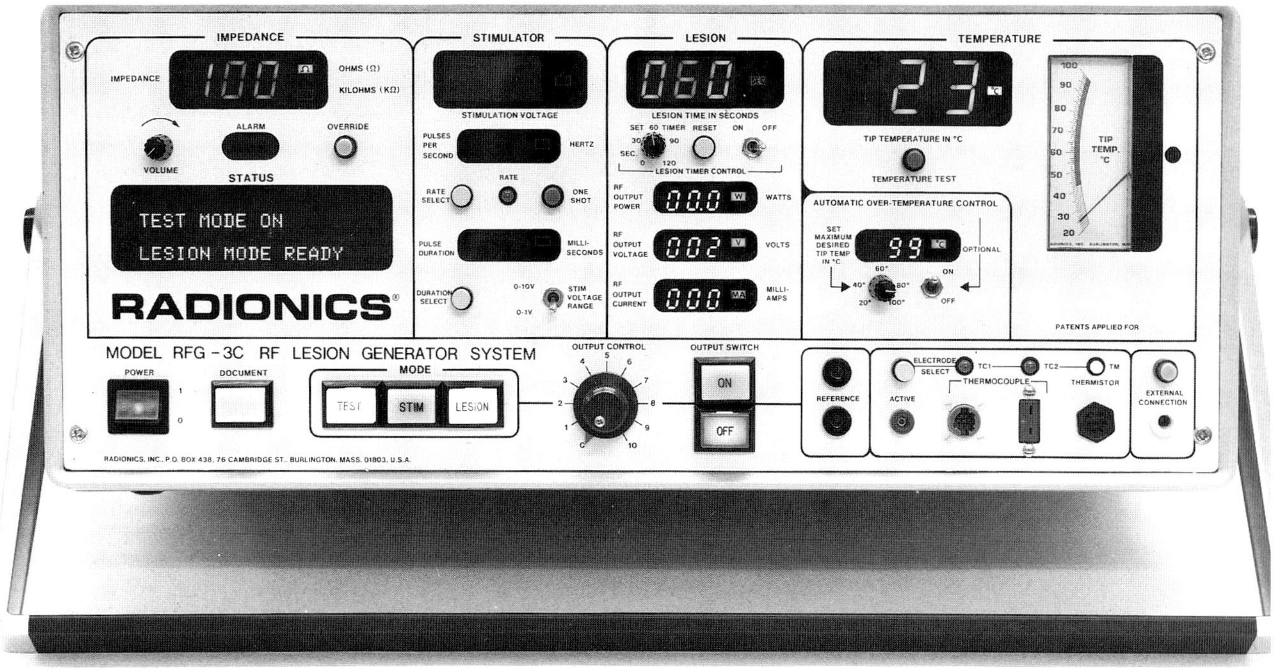

Figure 1. The Radionics, radiofrequency lesion generator.

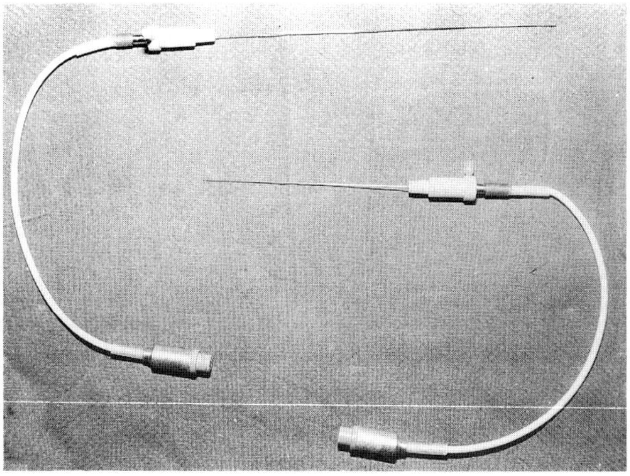

Figure 2. Radionics thermocouple electrodes.

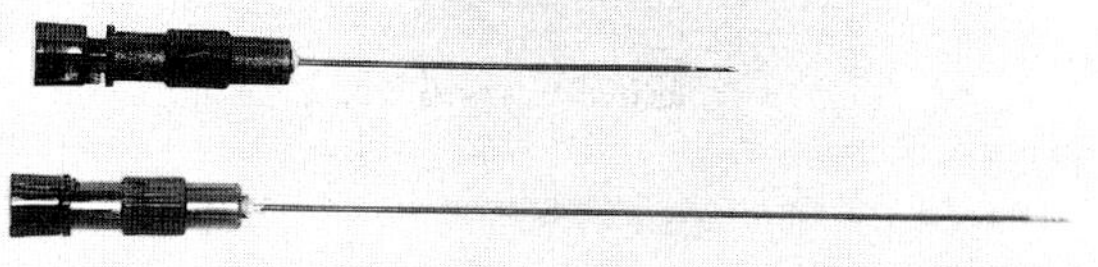

Figure 3. Sluijter–Mehta cannulae.

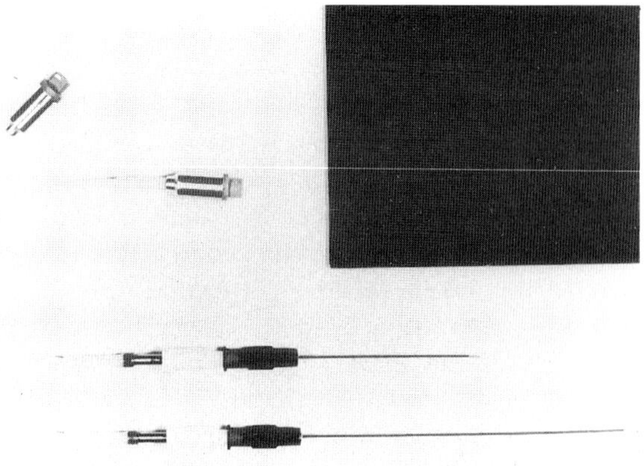

Figure 4. Sluijter–Mehta cannulae with thermocouple electrodes inserted.

of current only just maintains the temperature, and at the edge of the lesion no further spread takes place, owing to the natural dispersive effect of normal tissue. The further the distance from the cannula tip, the weaker the current and therefore the less the heat generated in the tissues. At a temperature of about 44 °C no permanent neurological damage will occur; thus, for practical purposes, when we talk about lesion size, we mean the volume of tissue within the 44°C isotherm.

It is important to bear in mind that the needle itself does not heat up as the current flows through it. The heating effect is manifested solely in the tissues, being generated by the current flowing from the cannula: it does not occur by direct conduction from the hot cannula. Indeed, the cannula actually acts as a heat-sink

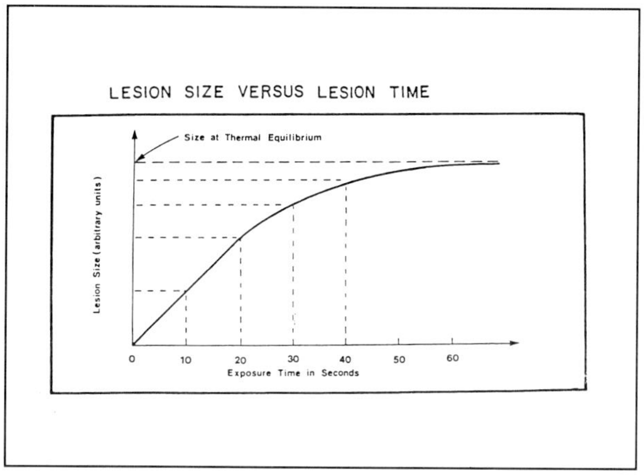

Figure 5.

absorbing some of the heat generated; this causes a small decrease in the lesion temperature next to the needle surface. The thermocouple probe records the temperature in the tissue being lesioned, but because of this slight drop in temperature next to the probe (sited, of course, within the cannula) the highest temperature generated in the tissues is slightly higher than that recorded by the thermocouple. In general, a meter reading of 85 °C should be regarded as the highest temperature which can reasonably be used; much higher temperatures cause boiling and drying and can even reduce the size of the lesion.

If, at a given probe temperature, lesion size is plotted against exposure time, it will be observed that the build-up of the lesion with time is relatively linear over the early part of the curve but then reaches the steady state situation referred to above (Figure 5). Thus, when a temperature-controlled lesion is being made, the voltage should be turned up until the desired temperature is achieved and then held at that temperature for 60 s. This ensures that the lesion has reached its maximum spread for that given temperature. Current thinking holds that there is no particular advantage is lesioning for longer than 60 s when performing facet joint denervation or rhizolysis; however, recent work carried out at Guy's Hospital suggests that 120 s would be a more appropriate lesioning time in order to obtain complete neurodestruction (2). Throughout the text, when discussing radiofrequency, where we have given the lesioning time as 60 s we would ask the reader to bear this latest information in mind when carrying out the procedures in question.

Some radiofrequency lesion-making machines do not have temperature-monitoring facilities. With these machines you cannot use the Sluijter–Mehta radiofrequency cannula; in this instance, you use instead the pole needle (Figure 6). This needle does not have a central removable stylet. Instead of noting temperature for the purposes of lesion-making, you must note instead the voltage level. With most machines, 22 V corresponds to a temperature of about 80 °C and 24 V corresponds to a temperature of 85 °C but the exact settings will depend on the individual machine in use.

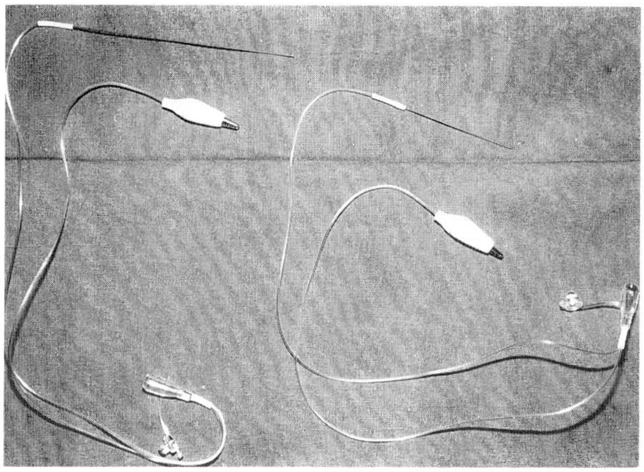

Figure 6. Pole needles.

Impedance

Most lesion generators have an impedance meter built into them. This allows the operator to monitor the impedance at the cannula tip. This facility is of special use when a percutaneous cordotomy is being performed; thus, the impedance will vary from about 400 Ω to 600 Ω in the extradural tissue, drop down to about 200 Ω in the cerebrospinal fluid and rise to about 700 Ω in the cord itself.

The impedance signal can be converted to a varying audible pitch by the lesion generator, which allows the various tissue interfaces to be 'heard' while the operator concentrates on the procedure. This obviates the need to watch the meter readings. Impedance readings are not really of much value when peripheral lesions are being carried out, except that if outside the 400–600 Ω range, a faulty connection in the thermocouple probe or cable can be suspected. A low-resistance shunting is experienced when cerebrospinal fluid is close to the lesion site. This factor should be particularly borne in mind when lesioning the trigeminal ganglion. The low impedance of the cerebrospinal fluid tends to 'drag' the current towards it, thus reducing the size of the lesion elsewhere.

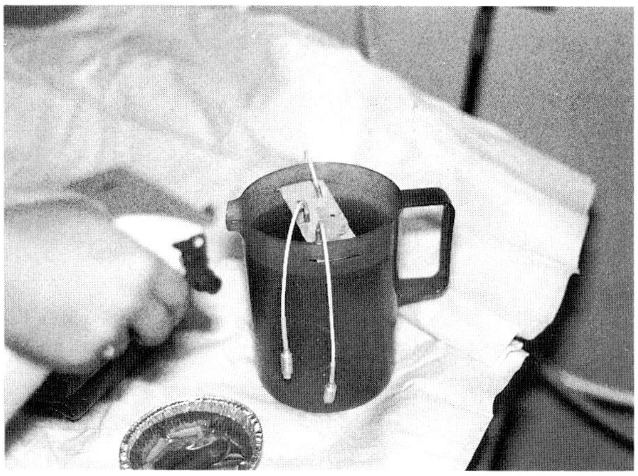

Figure 7. Sterilising the radiofrequency electrode with Cidex.

1. The radiofrequency probe is held in a container of sterilizing solution, e.g. Gigisept or Cidex for 30 minutes. The exposed part of the probe together with its lead is liberally sprayed with alcohol. The probe 'collar' is immersed in a gallipot full of sterilizing solution for 30 minutes (Figure 7).

2. The long lead is thoroughly sprayed with alcohol (Figure 8).

3. The probe and its collar are rinsed in water to wash off the sterilizing solution, put together and then thoroughly sprayed with alcohol (Figure 9).

4. Probe and collar should be *thoroughly* cleaned immediately after use.

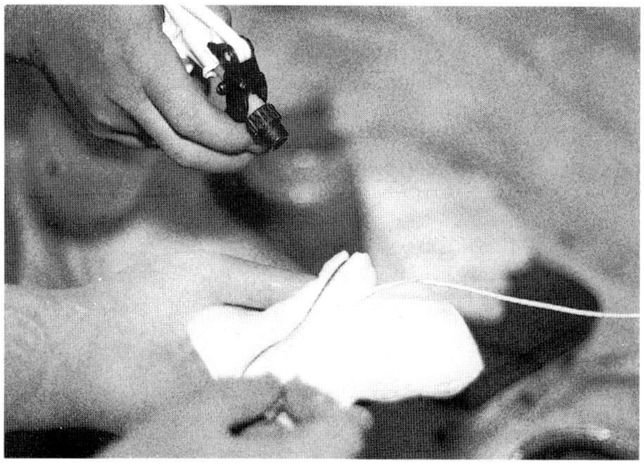

Figure 8. Spraying the lead with alcohol.

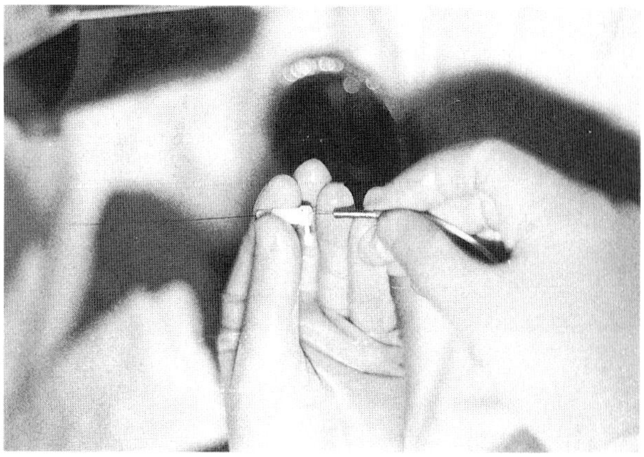

Figure 9. Collar fitted onto probe.

Soreness of Lesion Site

A practical point is that the area subjected to radiofrequency lesioning is often quite sore for a few weeks; recent work carried out at The Queen Elizabeth Military Hospital, Woolwich, suggests that the intensity and duration of this soreness may be reduced by injecting 1 ml of 0.5% Marcain plus 25 mg Hydrocortisone Acetate through the cannula after lesioning has been carried out at each site. (3)

Sterilization of Equipment

It is very important that all radiofrequency work be carried out under strict aseptic conditions. The Radionics equipment can be cleaned by low-temperature autoclaving (120 °C). If a low-temperature autoclave is not available, then the following technique can be used.

2 Cervical Facet Joint Denervation

Indication

Cervical facet joint pain, as confirmed by diagnostic injection – see page 91.

Clinical Picture

The patient presents with a stiff and painful neck, sometimes accompanied by a frank torticollis. The pain commonly radiates into the back of the head and sometimes over the shoulder and down the arm in a nonradicular fashion.

Pain is exacerbated by movement and is frequently positional. This often causes difficulty in finding a comfortable position in which to sleep. Pain is relieved by immobilization.

On examination cervical flexion is normally full and free. Extension, lateral rotation and lateral flexion are limited and painful. Muscle spasm is often present. Neurological examination is usually normal.

Technique

Sedation

Sedation is often necessary.

Position

The patient lies supine with his head on a radiolucent rest which affords full access to the neck; a C-arm image intensifier is positioned as shown in Figure 1.

Landmarks

Identify the mastoid process and mark its apex; draw a line from here down along the posterior border of the sternomastoid. Two fingers' breadth below the apex of the mastoid process corresponds approximately with the entry point for the C2/C3 facet joint.

View the cervical spine with the image intensifier positioned as shown in Figure 1; note that the intensifier end of the C-arm is situated to face the side being lesioned, while the X-ray tube part is situated on the opposite side. You now have a lateral view of the spine (Figure 2); rotate the C-arm so as to obtain an oblique view of the spine (Figure 3); in this view, the intervertebral foraminae and their associated facet joints are clearly seen (Figure 4).

Clean and drape the area.

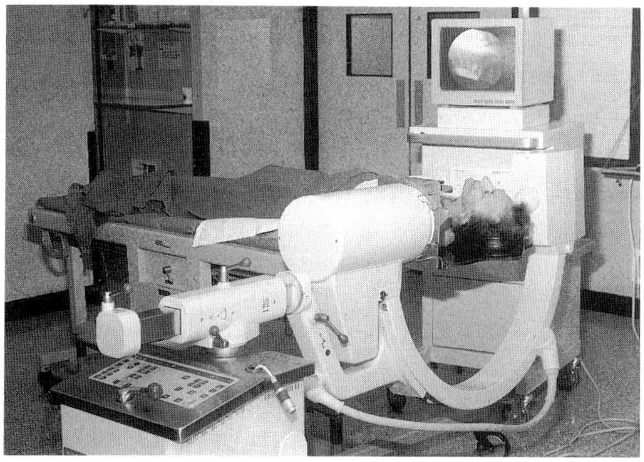

Figure 1. C-Arm positioned to give a lateral view of the cervical spine.

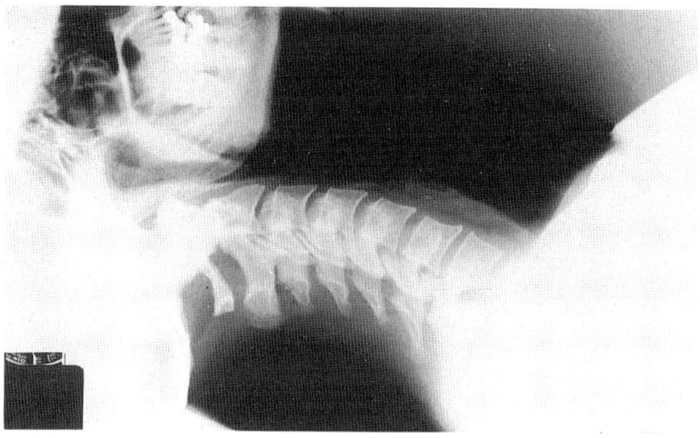

Figure 2. Lateral X-ray of the cervical spine.

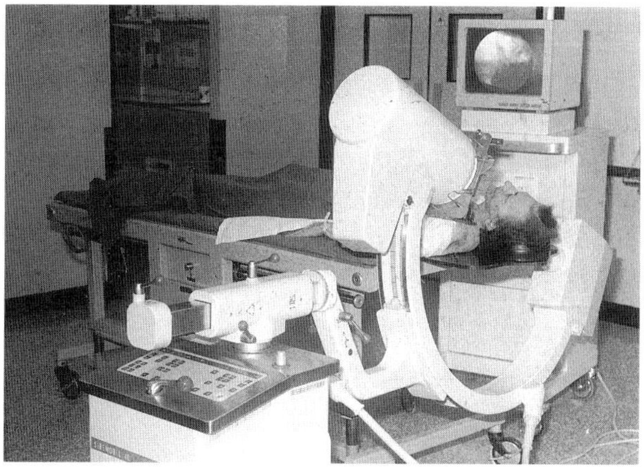

Figure 3. C-Arm positioned to give an oblique view of the cervical spine.

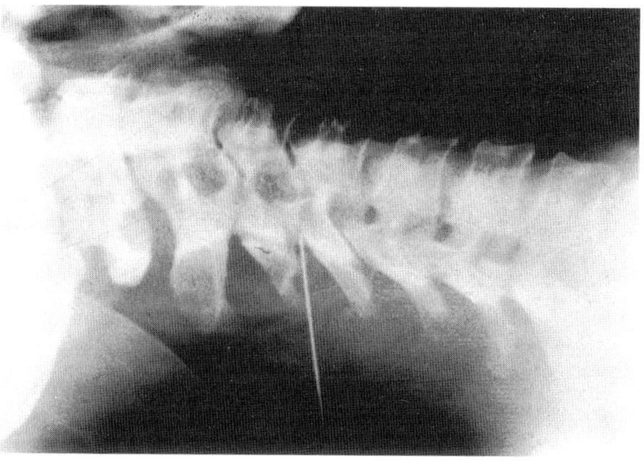

Figure 4. Oblique X-ray of the cervical spine showing the intervertebral foramina and their associated facet joints; the Sluijter–Mehta cannula is approaching its final position.

Procedure

- Infiltrate the entry point with 0.5% Lignocaine with Adrenaline (Figure 5).
- Insert the (50–60 mm) Sluijter–Mehta cannula or pole needle (Figure 6), aiming for the facet joint in question. The needle should be placed on the facet joint about 1–2 mm dorsal to the posterior (dorsal) edge of the intervertebral foramen (Figure 4). When you are happy with the position, if using a Sluijter–Mehta cannula, remove the stylet and insert the thermocouple probe (Figure 7).
- With the radiofrequency generator in 'stimulation' mode, try to elicit pain/strong paraesthesiae in the area of the patient's pain at 0.5 V or less and at a frequency of 100 Hz. When you have achieved this, change the frequency to 2 Hz and ensure that you do not elicit motor contractions in the root distribution of that level at less than twice the voltage necessary to achieve sensory stimulation.
- When you are satisfied with the position of your cannula, move on to the facet below. The entry point for your needle should be about one finger's breadth below the entry point above. We recommend that no more than three cervical facets be lesioned at one sitting. With more than this number, patient co-operation starts to wane and more sedation may be needed; heavy sedation tends to interfere with the patient's interpretation of the radiofrequency stimulation, which makes the procedure more difficult (Figure 8).
- When all the cannulae are correctly positioned, inject of 1 ml of 2% Lignocaine with Adrenaline into each in turn.

  If using Sluijter–Mehta cannulae, reinsert the thermocouple probe into the top cannula, wait for 1–2 min and then lesion at 80 °C for 60 s. Repeat the procedure for the other cannulae, removing each in turn when the lesioning is completed. We recommend that you increase the current gently, allowing the temperature to increase, and when it reaches 75 °C, activate the timing switch, which is

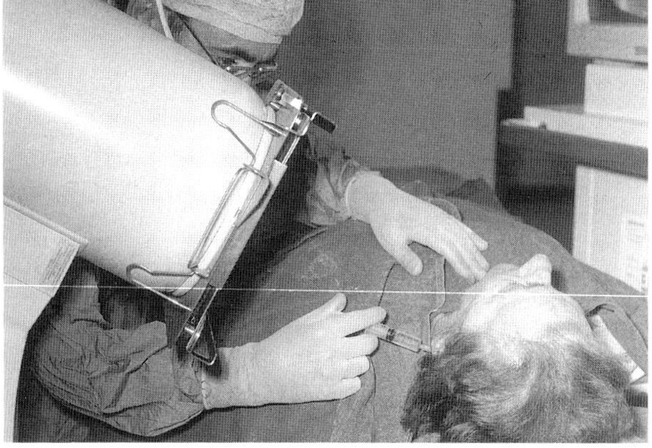

Figure 5. Infiltration with local anaesthetic along the axis of the X-ray tube.

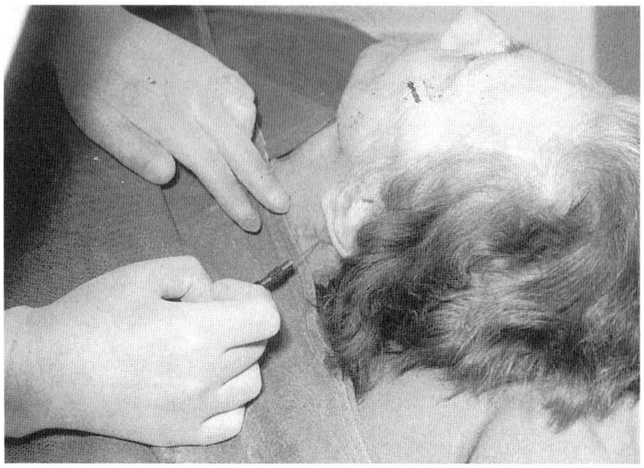

Figure 6. Inserting Sluijter–Mehta cannula.

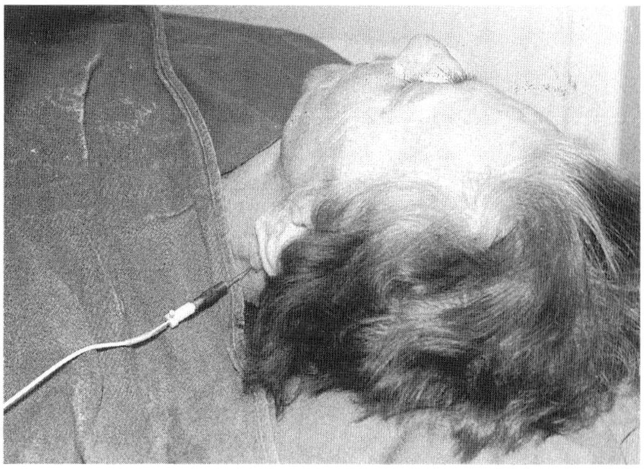

Figure 7. Sluijter–Mehta cannula in position with the thermocouple probe inserted.

> Note that when dealing with the C2/C3 facet joint, you have to deal with the branches of the posterior primary ramus of C2 to that joint. The way to do this is to position the cannula/pole needle on the vertebral arch of C2 at the level of the superior border of the C3 intervertebral foramen (Figure 9).

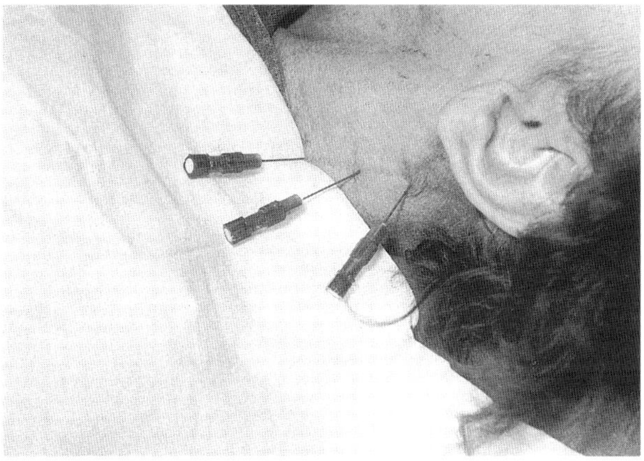

Figure 8. Three Sluijter–Mehta cannulae in position.

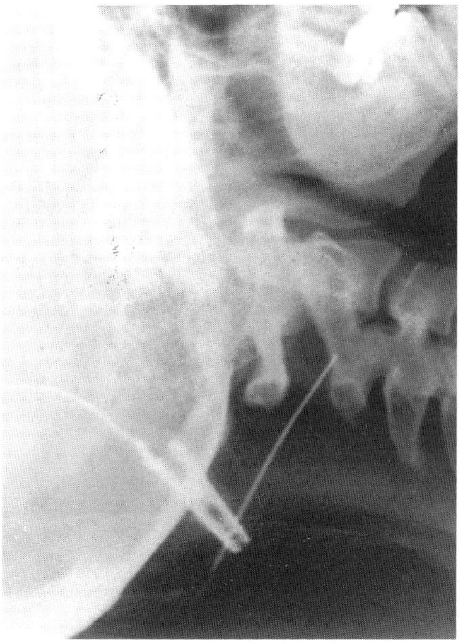

Figure 9. Pole needle on the vertebral arch of C2.

preset to 60 s. By the time this is done, the temperature will have reached 80 °C and should then be kept at that level.

If you are using a pole needle, lesion at 24 V for 60 s.

After-care

It is prudent to keep the patient in hospital for at least 30 min after completing the procedure.

Complications

Immediate

- Aspiration of blood: if this happens, reposition the cannula/pole needle.
- Aspiration of cerebrospinal fluid: this should not happen if your cannula/pole needle is always aimed at the target zones specified above.

Beware of intrathecal injections.

Late

General soreness in the area of the denervations with or without protective muscle spasm: this may last for a few weeks, so **warn the patient!**

3 Thoracic Facet Joint Denervation

Indication

Thoracic facet joint pain as confirmed by diagnostic injection–see page 93.

Clinical Picture

The patient presents with diffuse pain over the thoracic spine; the pain is often unilateral, radiating laterally over the thorax, and there may be muscle spasm present. The patient finds it uncomfortable to lie back in bed or to sit in a chair. Minimal activity exacerbates the pain.

On examination there is often some protective muscle spasm. As a rule, no clinical signs are present, except for tenderness in the paravertebral space overlying the affected facets.

Technique

Sedation

Sedation is often necessary.

Position

The patient lies prone on a radiolucent table, with the C-arm image intensifier positioned as shown in (Figure 1).

Landmarks

Each facet joint presents two primary targets which need to be identified by using the image intensifier. These are the superior and inferior poles of the facet joint in question (Figure 2a, b). Mark the skin overlying each spot with a cross.

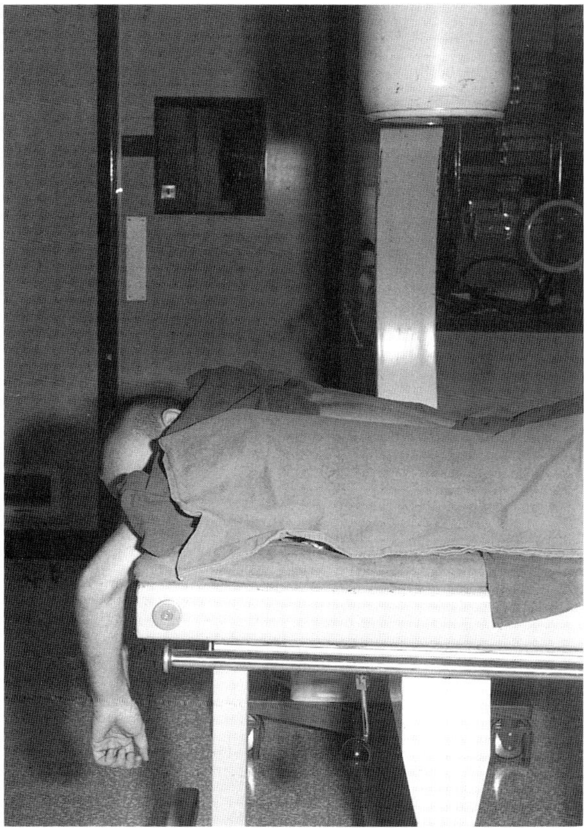

Figure 1. Patient positioned for thoracic facet joint denervation.

> **Entry Points**
> The safest way of reaching these targets is the MacEwan method, where the entry point is vertically above the target zone.

The area must be cleaned and draped.

Procedure

- Infiltrate the entry points with 0.5% Lignocaine with Adrenaline.
- Insert the 100 mm Sluijter–Mehta cannula or pole needle to reach the target zone. When you are happy with the position, if using a Sluijter–Mehta cannula, remove the stylet and insert the thermocouple probe.
- With the radiofrequency generator in 'stimulation mode', try to elicit pain/strong paraesthesiae at 0.5 V or less, at a frequency of 100 Hz. When you have achieved this, change the frequency to 2 Hz and ensure that you do not elicit motor contractions in the root distribution of that level at less than twice the voltage necessary to achieve sensory stimulation.

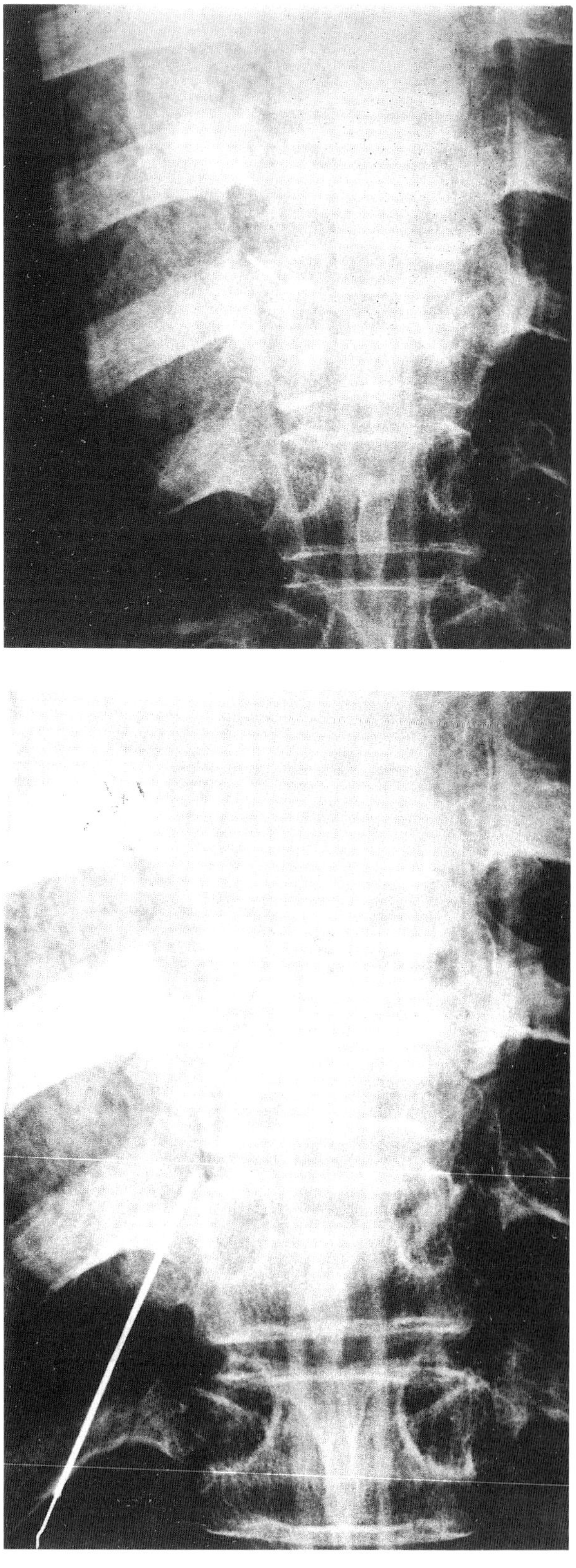

Figure 2.
a) Needle inserted onto superior pole of facet joint.
b) Needle inserted onto inferior pole of facet joint.

- When you are satisfied with the position of your cannula/pole needle, if using a Sluijter-Mehta cannula, remove the thermocouple probe and inject 1 ml of 2% Lignocaine with Adrenaline. Reinsert the thermocouple probe, wait for 1–2 min and then lesion at 85 °C for 60 s. We recommend that you increase the current gently, allowing the temperature to increase, and when it reaches 80 °C, activate the timing switch, which is preset to 60 s. By the time this is done, the temperature will have reached 85 °C and should then be kept at that level.

  If using a pole needle, lesion at 24 V for 60 s.

After-care

It is prudent to keep the patient in hospital for at least 30 min after completing the procedure.

Complications

Immediate

- Aspiration of blood: if this happens, reposition the cannula/pole needle.
- Aspiration of cerebrospinal fluid: this should not happen if your cannula/pole needle is always aimed at the target zones specified above.
- Pneumothorax: totally avoidable if you keep your cannula/ pole needle in contact with bone.

Late

General soreness in the area of the denervations with or without protective muscle spasm: this may last for a few weeks, so **warn the patient!**

Note

Bogduk *et al.* (1987) (4) call into question the efficacy of an electrode placed perpendicular to the nerve as outlined in our conventional approach above. They believe that there may be an inactive zone at the tip of the probe which may result in part of the nerve escaping thermocoagulation. They therefore recommend a modified approach where the electrode is introduced from below parallel to the target zone; they maintain that this creates a more radially spread lesion which is more effective.

We do not recommend this technique for inexperienced workers, since the safety factor of 'probe on bone' is removed. More experienced workers may find it of value in cases which do not respond to conventional radiofrequency lesioning as described above.

4 Lumbar Facet Joint Denervation

Indication

Lumbar facet joint pain as confirmed by diagnostic injection–see page 94.

Clinical Picture

The patient presents with a stiff and painful back. The back pain is often accompanied by pain in the buttocks and thighs, but does not normally extend below the knees, although it can do so. The pain is exacerbated by prolonged sitting and standing, and the patient normally finds it very difficult to get a good night's sleep because of the pain. Coughing and sneezing do not usually exacerbate the pain. The patient prefers to be mobile rather than still.

On examination there is often much protective muscle spasm. Forward flexion of the lumbar spine is usually full and free but it may sometimes, be restricted. Extension is invariably limited by sharp pain. Lateral flexion and lateral rotation may also be limited by pain. The facet joints are very tender to palpation. Straight leg raising is usually normal but may be accompanied by pain in the lower back. There are no root signs in pure facet joint pain. Neurological examination is normal.

Lumbar x-rays and CT scans are usually normal or perhaps show varying degrees of facet joint degeneration, but there is usually no correlation between the radiological appearance and the presence or intensity of the pain. There may be narrowing of the intervertebral disc space at the painful level.

If the patient has had a lumbar fusion, it is not at all uncommon for him or her to develop facet joint pain above the fusion level. Patients who have had disc material removed surgically may have loss of disc height, causing increasing pressure on the facet joints at that level.

Technique

Sedation

Sedation is often necessary.

Position

The patient lies prone with the C-arm image intensifier positioned for screening in the anteroposterior plane as shown in Figure 1.

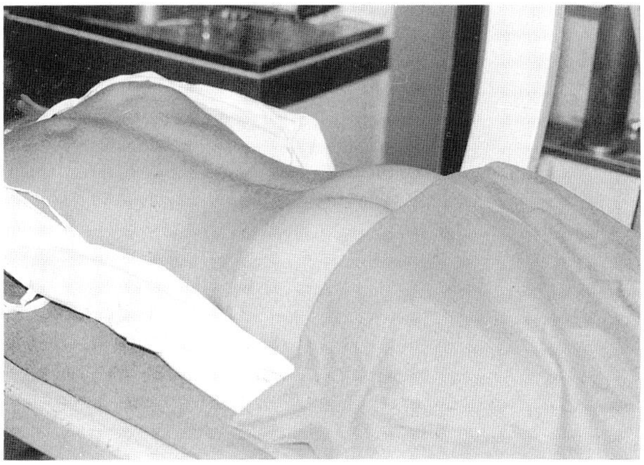

Figure 1. Patient positioned for lumbar facet joint denervation.

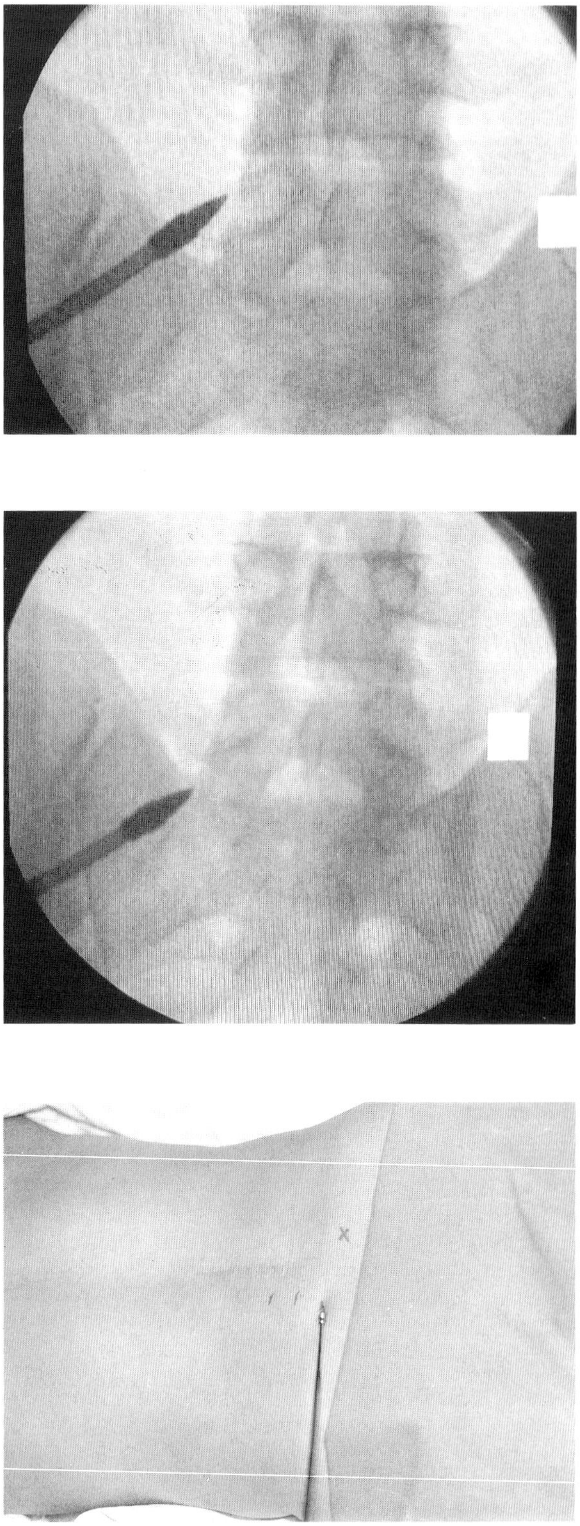

Figure 2.
a) Surface marker overlying target point on transverse process of L5.
b) Surface marker overlying target point in sacral groove.
c) Surface marking of entry points (L4, L5, S1).

Landmarks

Each facet joint presents two primary targets which need to be identified by using the image intensifier: these are the junctions of transverse processes and laminae above and below the joint in question (Figure 2a). In the case of the lumbosacral facet joints, the corresponding target is situated just below the junction of the lateral mass of the sacrum and its superior articular process (Figure 2b). The skin overlying the entry points can be marked for easy reference (Figure 2c).

Other additional targets may also be used (Figure 3); some workers use all the targets shown in Figure 3 routinely, while others use only the primary targets and reserve the additional targets for recalcitrant cases.

Entry Points

There are two ways of reaching these targets:

- The entry point is at the tip of the transverse process and the needle is angled medially to the target zone. This method is the preferred method if one is going for multiple targets.
- The entry point is directly above the target zone. This method is ideal for those using primary targets only.

The area must be cleaned and draped.

Procedure

- Infiltrate the entry points with 0.5% Lignocaine with Adrenaline.
- Insert the 100 mm Sluijter–Mehta cannula or pole

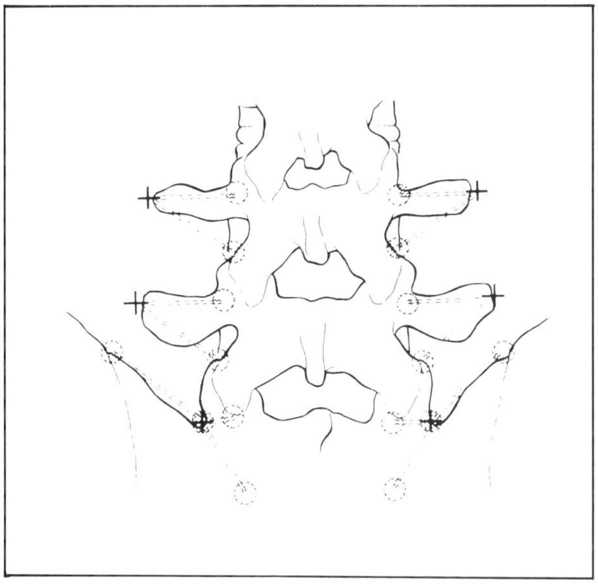

Figure 3. Target points for L4/5 and L5/S1 facet joints, the crosses indicate the skin entry points.

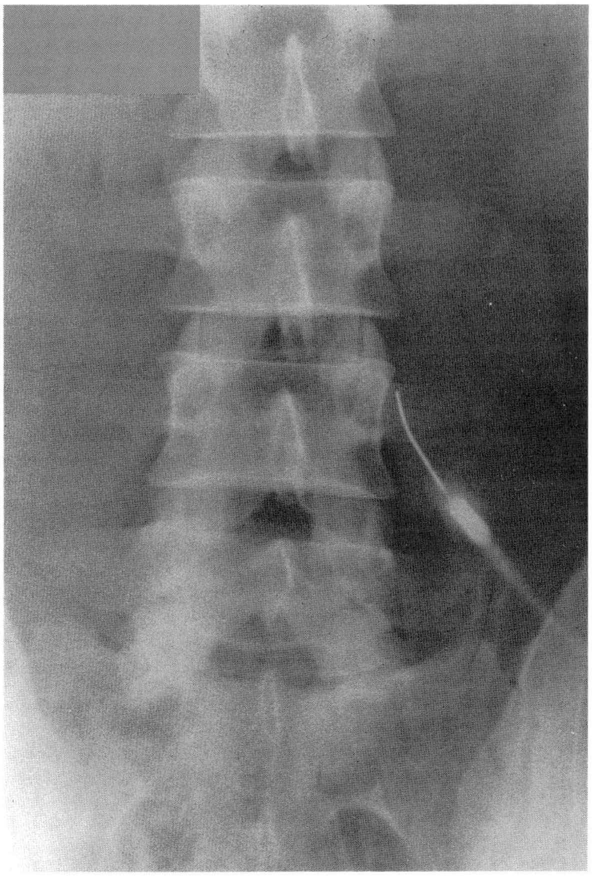

Figure 4. Sluijter–Mehta cannula inserted onto target point on L4 transverse process.

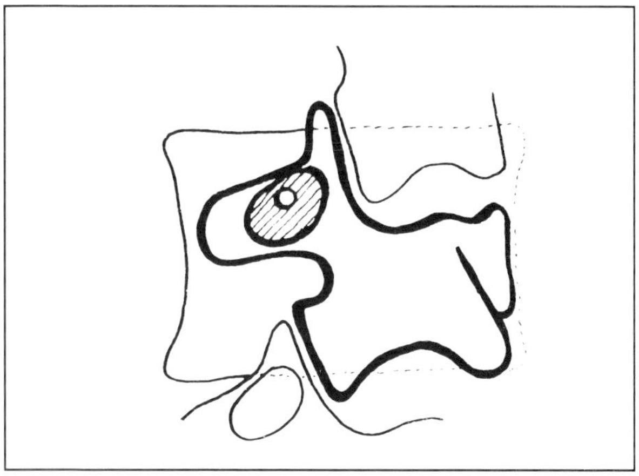

Figure 5. The eye of the 'Scotty dog'.

needle to reach the target zone (Figure 4). The correct position of the cannula/pole needle can be checked by radiological screening with the C-arm in an oblique plane at 45 °C off the vertical towards the side being treated. In this view the tip of the cannula/pole needle will be in the 'eye' of the classical 'Scottie dog' picture (Figure 5). When you are happy with the position, if using a Sluijter–Mehta cannula, remove the stylet and insert the thermocouple probe.

- Try to elicit pain/strong paraesthesiae with the radiofrequency generator on 'stimulation mode' at 0.5 V or less at a frequency of 100 Hz. When you have achieved this, change the frequency to 2 Hz and ensure that you do not elicit motor contractions in the root distribution of that level at less than twice the voltage necessary to achieve sensory stimulation.
- When you are satisfied with the position of your cannula/pole needle, if using a Sluijter–Mehta cannula, remove the thermocouple probe and inject 1 ml of 2% lignocaine with adrenaline. Reinsert the thermocouple probe, wait for 1–2 min and then lesion at 85 °C for 60 s. We recommend that you increase the current gently, allowing the temperature to increase, and when it reaches 80 °C, activate the timing switch, which is preset to 60 s; by the time this is done, the temperature will have reached 85 °C and should then be kept at that level.

If using a pole needle, lesion at 24 V for 60 s.

After-care

It is prudent to keep the patient in hospital for at least 30 min after completing the procedure.

Complications

Immediate

- Aspiration of blood: if this happens, reposition the cannula/pole needle.
- Aspiration of cerebrospinal fluid: this should not happen if your cannula/pole needle is always aimed at the target zones specified above.

Late

General soreness in the area of the denervations with or without protective muscle spasm: this may last for a few weeks, so **warn the patient!**

> **Note**
> Bogduk *et al.* (1987) (4) call into question the efficacy of an electrode placed perpendicular to the nerve as outlined in our conventional approach above. They believe that there may be an inactive zone at the tip of the probe which may result in part of the nerve escaping thermocoagulation. They therefore recommend a modified approach where the electrode is introduced from below parallel to the target zone; they maintain that this creates a more radially spread lesion which is more effective.
>
> We do not recommend this technique for inexperienced workers, since the safety factor of 'probe on bone' is removed. More experienced workers may find it of value in cases which do not respond to conventional radiofrequency lesioning as described above.

5 Sacroiliac Joint Denervation

Indication

Sacroiliac joint pain as confirmed by diagnostic injections and for which repeated injections of local anaesthetic and steroid into the sacroiliac joint have failed to provide permanent relief.

Clinical Picture

The patient with sacroiliac joint pain presents with a stiff and painful back. The back pain is often accompanied by pain in the buttocks and thighs, but does not normally extend below the knees, although it can do. The pain is aggravated by prolonged sitting and standing, and the patient normally finds it very difficult to get a good night's sleep. Coughing and sneezing do not usually exacerbate the pain. The patient prefers being mobile to being still. This condition is sometimes difficult to distinguish from L5/S1 facet joint pain but you can usually elicit tenderness over the sacroiliac joints.

On examination flexion of the lumbar spine is usually full and free but it may sometimes be restricted. Extension is invariably limited by sharp pain. The sacroiliac joints are very tender to palpation; the straight leg raising test is usually normal but may be accompanied by pain in the lower back. There are no root signs and neurological examination is normal. Lumbar x-rays may be of value, as may bone scans (sacroiliitis). Sacroiliac joint injections would have been tried (see page 95).

Technique

The technique is virtually identical with that for lumbar facet joint denervation: in fact, the first step of sacroiliac joint denervation is to carry out L4/L5 and L5/S1 facet denervations (see page 9). In addition to the points which are targeted for L4/L5 and L5/S1 facet denervations, you will also need to target the following point (Figure 1):

- Contribution from S1 to the sacroiliac joint, located at a point just lateral to the S1 foramen;
- Contribution from S2 to the sacroiliac joint, located at a point just lateral to the S2 foramen;
- The superior and inferior poles and the central point of the sacroiliac joint.

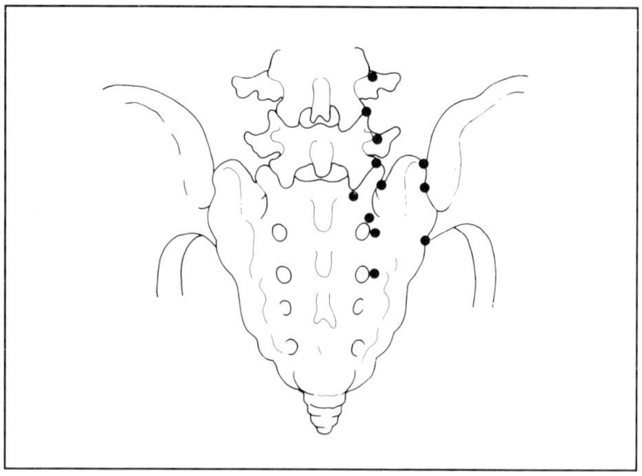

Figure 1. Target points for radiofrequency denervation of sacroiliac joint.

The procedure is not always successful, owing to the very rich nerve supply of the sacroiliac joint, and may need to be repeated.

6 Cervical Rhizolysis

Indication

Monoradicular cervical root pain in the C2–C4 dermatomes, confirmed by diagnostic injection (see page 79). The pain is not responsive to other more conservative therapy. We do not recommend that you carry out this technique below the C4 root, since numbness inevitably accompanies rhizolysis; numbness in the upper limb will be unacceptable to most patients.

Clinical Picture

Pain along a specific cervical spinal dermatome.

Technique

Sedation

Sedation is often necessary.

Position

The patient lies supine with the head on a radiolucent rest which affords full access to the neck; a C-arm image intensifier is positioned as shown (Figure 1).

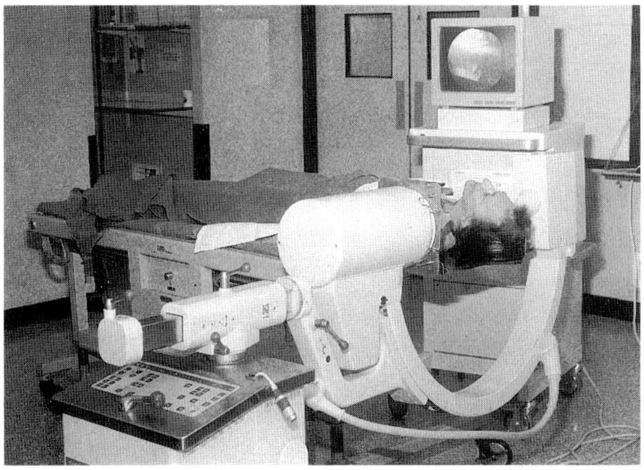

Figure 1. Patient positioned for cervical rhizolysis.

Landmarks

Identify the mastoid process and mark its apex; draw a line from here down along the posterior border of the sternomastoid. Two fingers' breadth below the apex of the mastoid process corresponds approximately with the entry point for the C2/C3 intervertebral foramen; each subsequent entry point level is about one finger's breadth further down.

View the cervical spine with the image intensifier positioned as shown in Figure 1; note that the intensifier end of the C-arm is situated to face the side being lesioned, while the x-ray tube part is situated on the opposite side. You now have a lateral view of the spine (Figure 2). Rotate the C-arm so as to obtain an oblique view of the spine (Figure 3); in this view, the intervertebral foramina and their associated facet joints are clearly seen (Figure 4).

Clean and drape the area.

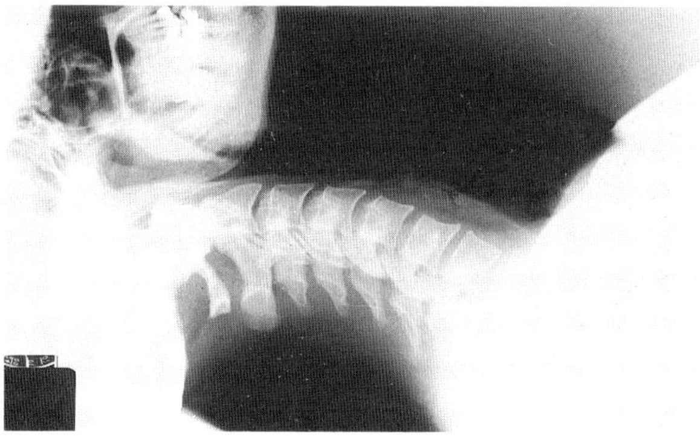

Figure 2. Lateral X-ray of cervical spine.

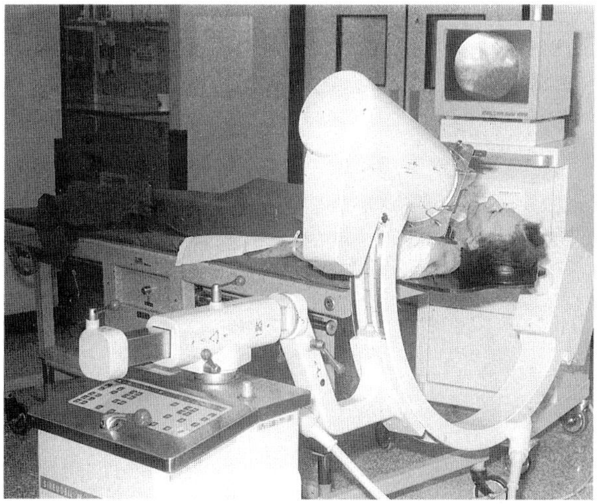

Figure 3. X-ray tube angled to give an oblique view of the cervical spine.

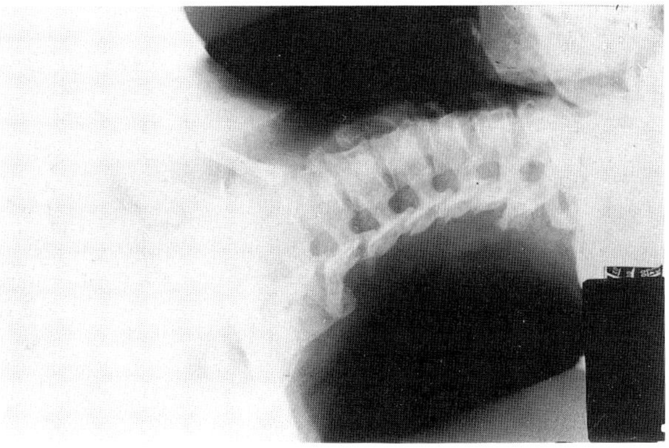

Figure 4. Oblique view of the cervical spine to show the intervertebral foramina.

Procedure

- Infiltrate the entry point with 0.5% Lignocaine with Adrenaline.
- Insert the (50–60 mm) Sluijter–Mehta or pole needle, aiming for the intervertebral foramen in question. The cannula should be placed on bone at the posterior (dorsal) margin of the foramen and then 'walked' just into the foramen, ensuring that it remains at the posterior part of the foramen (Figures 5, 6).

The C2 root

Rhizolysis of the C2 nerve root involves a technique which is slightly different from the one just described. The C2 nerve root does not occupy a proper intervertebral foramen but rather the space between the vertebral arch of C1 and C2. This space is best visualised with the C-arm image intensifier in the lateral axis. It is not unlike a house with straight 'walls' situated cephalad and caudally and sloping 'roofs' situated ventrally. Your needle should be inserted so as to strike the vertebral arch of C1 or C2 at the junction of the cephalad 'wall' with the 'roof' (Figure 7a). The needle should then be lifted gently off the bone and positioned in the middle of an imaginary line joining the two roof-wall' junctions. The needle should now lie on the C2 nerve root (Figure 7b).

With all roots

Check the anteroposterior view so as to ensure that you have not advanced the cannula/needle too far into the foramen; in this view, the cannula/ needle tip should be no more medial than an imaginary line drawn through the facet joints (Figure 8). When you are satisfied with the position, if using a Sluijter–Mehta cannula, remove the stylet and insert the thermocouple probe.

- With, the radiofrequency generator in stimulation mode, try to elicit pain/strong paraesthesiae in the area of the patient's pain at 0.5 V or less and at a frequency of 100 Hz. When you have achieved this, change the frequency to 2 Hz and ensure that you do not elicit motor contractions in the root distribution of that level at less than twice the voltage necessary to achieve sensory stimulation.
- When the cannula/needle is correctly positioned, if using a Sluijter–Mehta cannula, remove the thermocouple probe and inject 1 ml of 2% Lignocaine through it. Reinsert the thermocouple probe, wait for 1–2 min and then lesion at 60 °C for 60 s. We recommend that you increase the current gently, allowing the temperature to increase, and when it reaches 55 °C, activate the timing switch, which is preset to 60 s; by the time this is done, the temperature will have reached 60 °C and should then be kept at that level.

  If using a pole needle, lesion at 20 V for 60 s (exact settings will vary with your machine).
- We strongly recommend that you lesion initially at

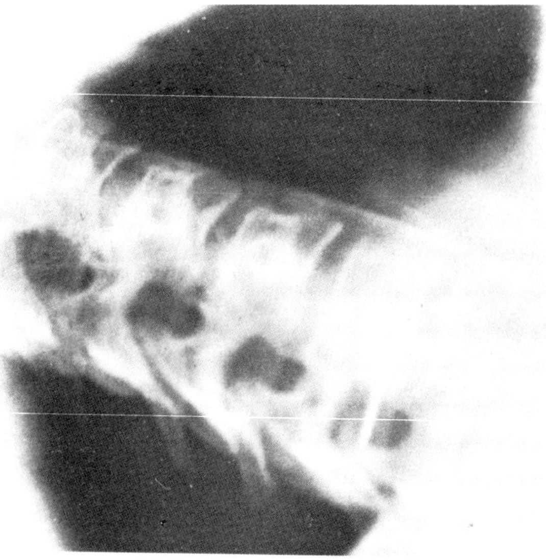

Figure 5. Needle in contact with bone at the posterior (dorsal) margin of the foramen.

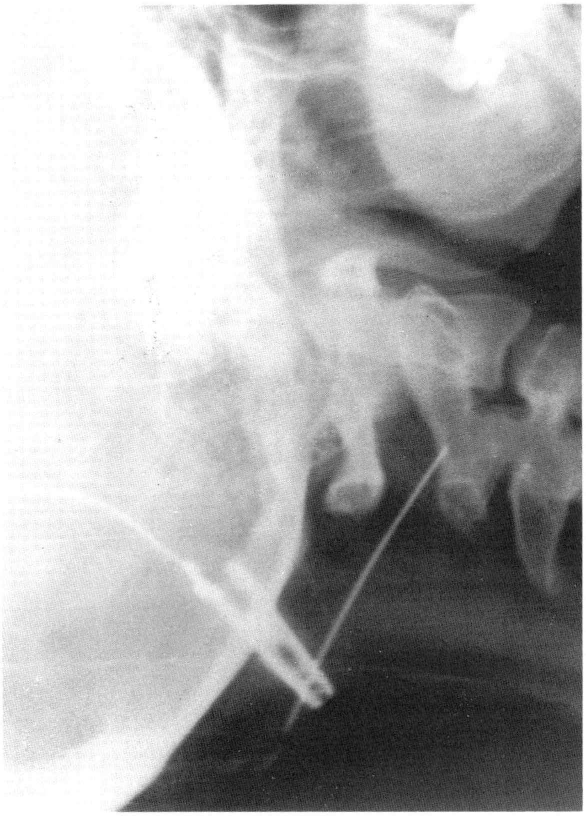

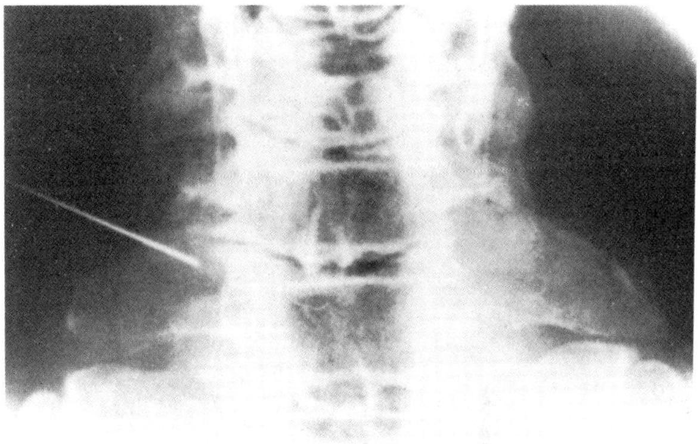

Figure 6. PA view needle in cervical foramen.

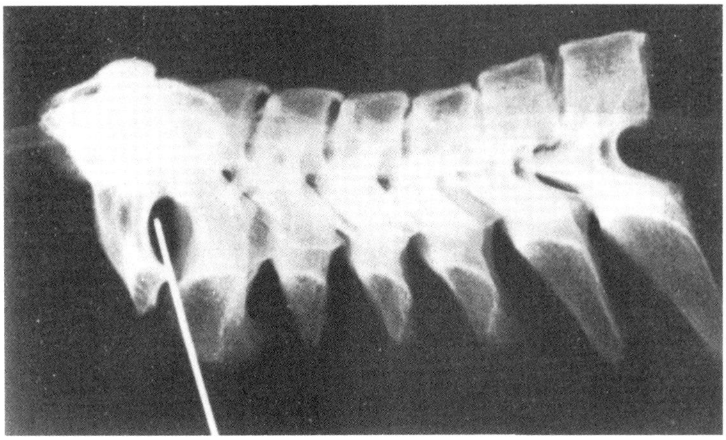

Figure 7.
a) Needle in contact with the vertebral arch of C2.
b) Needle positioned on the C2 nerve root.

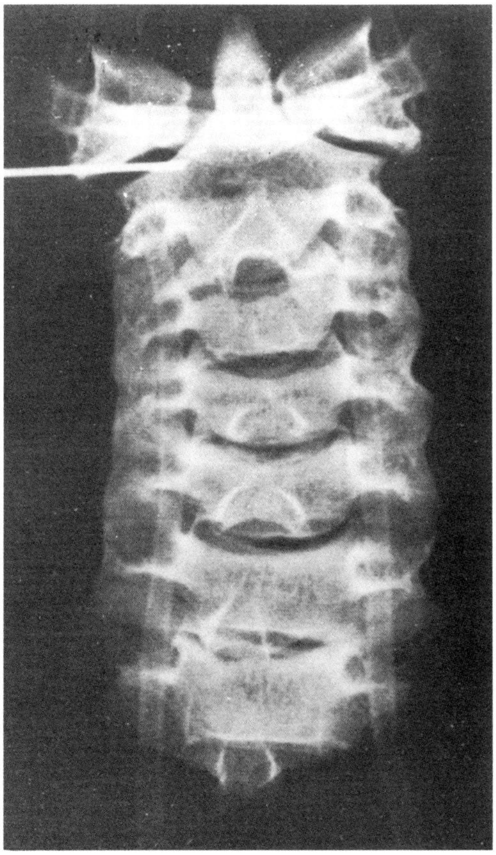

Figure 8.
PA view, needle positioned
for C2 rhizolysis.

a low temperature and, if necessary, repeat the procedure on several occasions at slightly higher temperatures in an effort to get a permanent effect. We would not recommend a lesioning temperature greater than 80 °C for fear of damaging the motor root.

After-care

It is prudent to keep the patient in hospital for at least 30 min after completing the procedure.

Complications

Immediate

- Aspiration of blood: if this happens, resite needle.
- Aspiration of cerebrospinal fluid: this is unlikely to happen if your needle is never advanced past an imaginary line joining the facet joints in the anteroposterior view on the image intensifier. **Beware of intrathecal injections!**
- Motor nerve damage: avoidable if this technique is followed.

Late

General soreness in the area of the denervations with or without protective muscle spasm. This may last for a few weeks, so **warn the patient!**

7 Thoracic Rhizolysis

Indication

Monoradicular thoracic root pain as confirmed by diagnostic injection (see page 81) and not responsive to other more conservative therapy. There is always the possibility of creating deafferentation pain, so rhizolysis must be considered as a last resort.

Clinical Picture

The patient presents with pain along a single thoracic nerve. The pain may have all or some of the characteristics of a neuralgia, i.e. burning pain with a shooting component, hypersensitivity, areas of patchy sensation, allodynia, paraesthesiae and dysaesthesiae.

Technique

Sedation

Sedation is often necessary.

Position

The patient lies prone on a radiolucent table. With upper thoracic work it is advisable to arrange the patient's arms so that they hang over the top edge of the table; this manoeuvre lifts the scapulae out of the way (Figure 1).

Landmarks

With the C-arm image intensifier positioned for screening in the anteroposterior plane, the level of the intervertebral foramen in question is located; the adjacent head of rib is then identified and a skin mark is made over the intercostal space, 5 cm from the midline, just below the caudal edge of this rib. This is your entry point.

Clean and drape the area.

Procedure

- Infiltrate the entry point with 0.5% Lignocaine with Adrenaline; direct the needle towards the intervertebral foramen and infiltrate further but do not place any anaesthetic in the intervertebral foramen itself.
- Insert the 100 mm Sluijter–Mehta cannula or pole needle through the entry point and direct it towards the vertebral lamina overlying the intervertebral

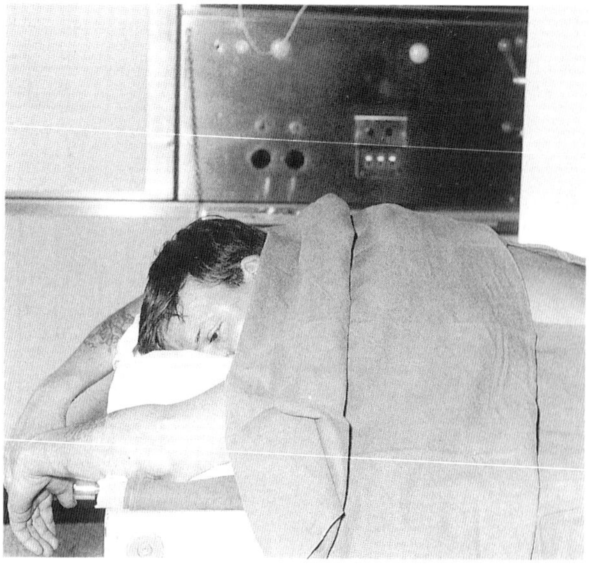

Figure 1. Patient positioned for thoracic rhizolysis.

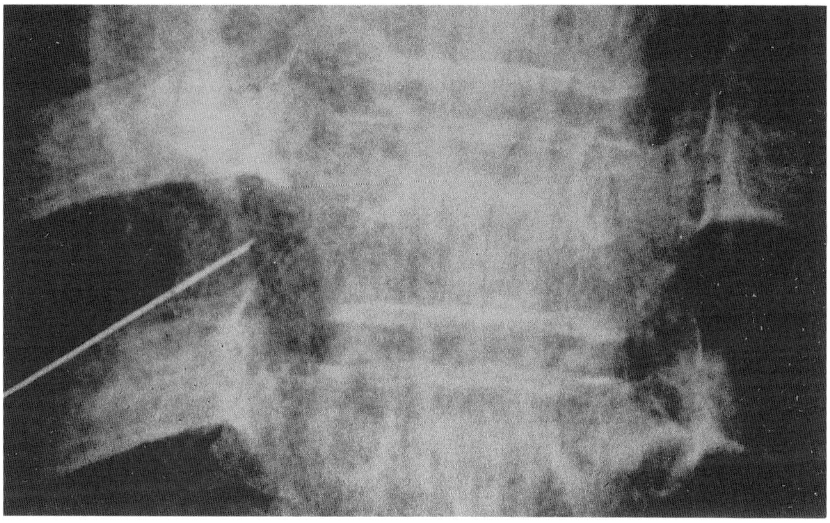

Figure 2. Pole needle in contact with vertebral lamina.

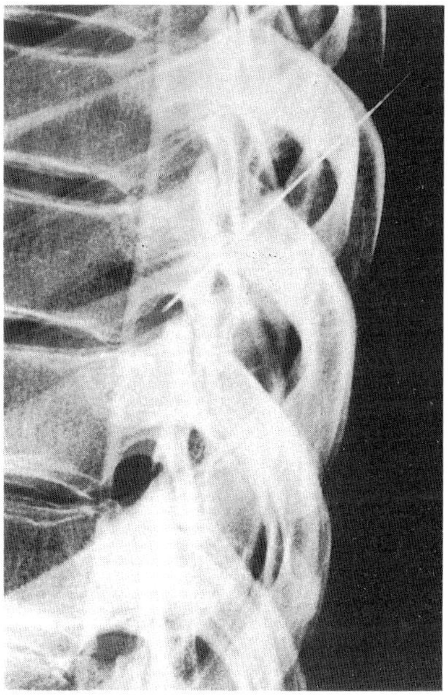

Figure 3. Pole needle in thoracic intervertebral foramen.

foramen (Figure 2). Walk the needle laterally off the lamina and into the foramen. You will feel a definite 'give' as the needle enters the foramen (Figure 3). Advance the needle into the foramen no more medial than an imaginary line joining the facet joints in the anteroposterior view obtained by the image intensifier.

- Turn the image intensifier so as to obtain a lateral view; it is not easy to obtain a true lateral view in the thoracic region as the ribs get in the way at all levels and the scapulae get in the way in the upper thoracic region. Angle the image intensifier tube so as to get the best lateral (or oblique) view possible. The needle should be lying in the posterior half of the foramen. Remember that the front half is motor territory; if necessary, pull the needle back into the posterior half of the foramen (Figure 3). When you are satisfied with the position, if using a Sluijter–Mehta cannula, remove the stylet and insert the thermocouple probe.
- With, the radiofrequency generator in stimulation mode, try to elicit pain/strong paraesthesiae in the area of the patient's pain at 0.5 V or less and at a frequency of 100 Hz. When you have achieved this, change the frequency to 2 Hz and ensure that you do not elicit motor contractions in the root distribution of that level at less than twice the voltage necessary to achieve sensory stimulation.
- When the needle is correctly positioned, if using a Sluijter–Mehta cannula, remove the thermocouple probe and inject 1 ml of 2% Lignocaine through it. Reinsert the thermocouple, wait for 1–2 min and then lesion at 70 °C for 60s. We recommend that you increase the current gently, allowing the temperature to increase, and when it reaches 65 °C, activate the timing switch, which is preset to 60 s; by the time this is done, the temperature will have reached 70 °C and should then be kept at that level.

  If using a pole needle, lesion at 22 V for 60 s (exact settings will vary with your machine).
- We strongly recommend that you lesion initially at a low temperature and repeat the procedure on several occasions at a slightly higher temperature in an effort to get a permanent effect. We would not recommend a lesioning temperature greater than 80 °C for fear of damaging the motor root.

After-care

It is prudent to keep the patient in hospital for at least 30 min after the lesioning; be on the look-out for the sudden onset of dyspnoea and/or chest pain, which may indicate a pneumothorax.

Complications

Immediate

- Aspiration of blood; if this happens, resite needle.
- Aspiration of cerebrospinal fluid: this is unlikely to happen if your needle is never advanced past an imaginary line joining the facet joints in the

anteroposterior view on the image intensifier. **Beware of intrathecal injections!**

- Motor nerve damage: avoidable if this technique is followed.
- Pneumothorax: always a possibility the risk of which can be minimised by meticulous technique.

Late

General soreness in the area of the denervations with or without protective muscle spasm. This may last for a few weeks, so **warn the patient!**

8 Lumbar Rhizolysis

Indication

Monoradicular sciatica confirmed by diagnostic injection (see page 83) and not responsive to more conservative therapy. There is always the possibility of creating deafferentation pain, so this technique must be regarded as a last resort; in addition, the numbness in the lower limb which inevitably accompanies the procedure may be unacceptable to the patient.

Clinical Picture

The patient presents with pain down a single nerve root; more often than not, the patient will have had previous back surgery and the surgeons are not keen to re-explore the back. Other more conservative measures will have failed to help.

Technique

Sedation

Sedation is often necessary.

Position

The patient lies prone on a radiolucent table with a pillow under the abdomen in order to straighten the spine (Figure 1).

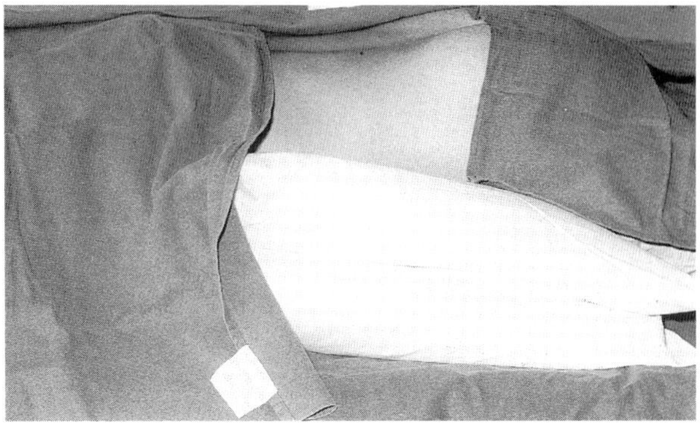

Figure 1. Patient positioned for lumbar rhizolysis.

Landmarks

With the C-arm image intensifier positioned for screening in the anteroposterior plane, the level of the intervertebral foramen in question is located; the adjacent transverse process is identified and a skin mark is made below the tip of the transverse process about 2 cm caudad to the intervertebral foramen (Figure 2). This mark is usually about 5 cm from the midline.

Clean and drape the area.

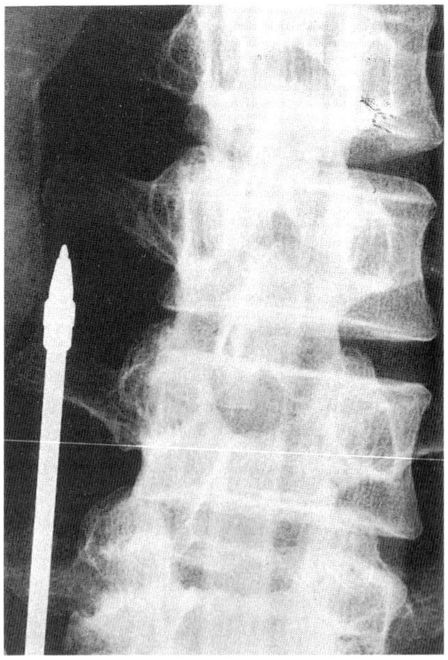

Figure 2. Surface marker defining skin entry point.

Procedure

- Infiltrate the entry point with 0.5% Lignocaine with Adrenaline; direct the needle towards the intervertebral foremen and infiltrate but do not place any anaesthetic in the intervertebral foramen itself.

- Insert the 100 mm Sluijter–Mehta cannula or pole needle through the entry point and direct it towards the vertebral lamina overlying the intervertebral foramen; 'walk' the needle laterally off the lamina and into the foramen. You will feel a definite 'give' as the needle enters the foramen; advance the needle into the foramen no more medial than an imaginary line joining the facet joints in the antero-posterior view obtained on the image intensifier (Figure 3).
- Rotate the image intensifier so as to obtain a lateral view; the cannula/needle tip should be lying in the posterior half of the intervertebral foramen. Remember that the front (ventral) half of the intervertebral foramen is motor territory; if necessary, pull the tip back into the posterior half of the intervertebral foramen (Figure 4). If you experience difficulty inserting the cannula/needle into the foramen, try a more lateral entry point. When you are satisfied with the position, if using a Sluijter–Mehta cannula, remove the stylet and insert the thermocouple probe.
- With the radiofrequency generator in stimulation mode, try to elicit pain/strong paraesthesiae in the area of the patient's pain at 0.5 V or less and at a frequency of 100 Hz. When you have achieved this, change the frequency to 2 Hz and ensure that you do not elicit motor contractions in the root distribution of that level at less than twice the voltage necessary to achieve sensory stimulation.
- You may want to confirm your final position by withdrawing the thermocouple from the cannula and then injecting about 0.2 cm^3 of a water-soluble contrast medium through it in order to demonstrate the dorsal root ganglion at the tip of your needle. You may need to inject several aliquots of medium until you can demonstrate the dorsal root ganglion. With a pole needle, you can inject directly through the needle (Figure 5).

When the cannula/needle is correctly positioned, if using a Sluijter–Mehta cannula, remove the thermocouple probe and inject 1 ml of 2% Lignocaine through it. Reinsert the thermocouple probe, wait for 1–2 min and then lesion at 70 °C for 60 s. We recommend that you increase the current gently, allowing the temperature to increase, and when it reaches 65 °C, activate the timing switch, which is preset to 60 s; by the time this is done, the temperature will have reached 70 °C and should then be kept at that level.

If using a pole needle, lesion at 22 V for 60 s (exact settings will vary with your machine).

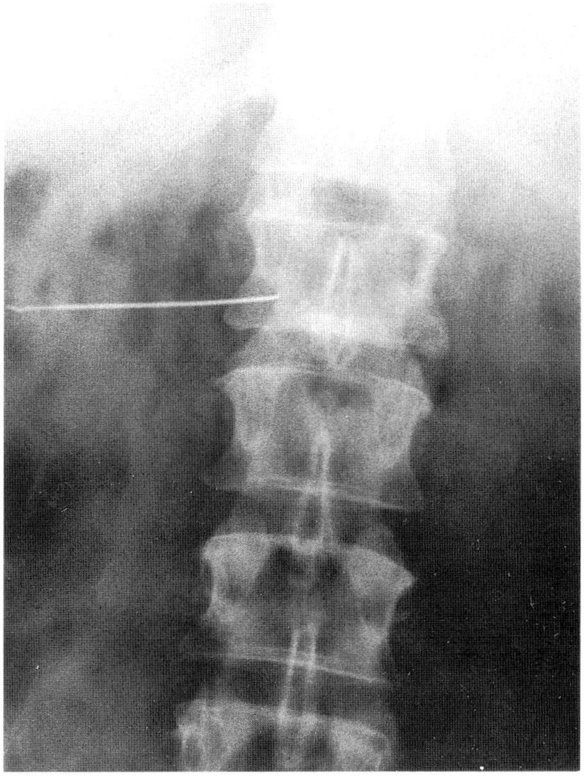

Figure 3. PA X-ray showing pole needle in foramen.

Figure 4. Lateral X-ray showing pole needle in foramen.

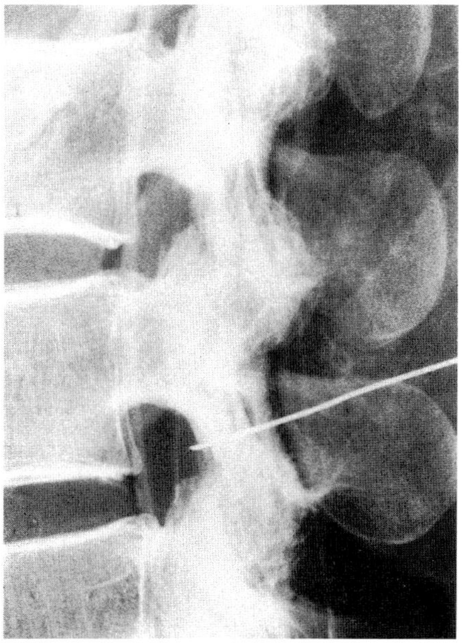

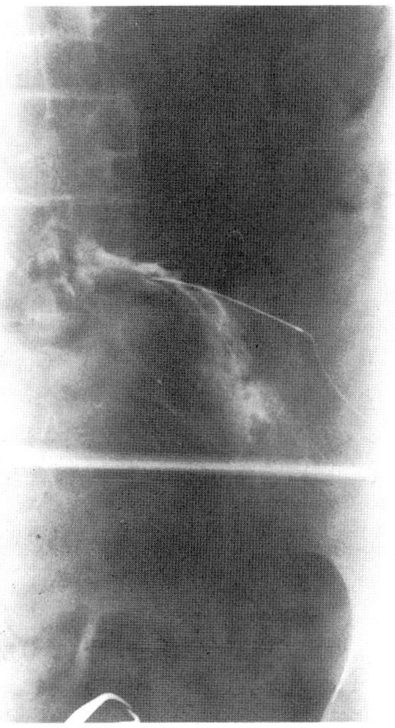

Figure 5. Pole needle in dorsal root ganglion.

- We strongly recommend that you lesion initially at a low temperature and repeat the procedure on several occasions at a slightly higher temperature in an effort to get a permanent effect. We would not recommend a lesioning temperature greater than 80 °C for fear of damaging the motor root.

After-care

It is prudent to keep the patient in hospital for at least 30 min after the lesioning;

Complications

Immediate

- Aspiration of blood: if this happens, resite the cannula/needle.
- Aspiration of cerebrospinal fluid; this is unlikely to happen if your cannula/needle is never advanced past an imaginary line joining the facet joints in the anteroposterior view on the image intensifier. **Beware of intrathecal injections!**
  - Motor nerve damage: avoidable if this technique is followed.

Late

General soreness in the area of the denervations with or without protective muscle spasm. This may last for a few weeks, so **warn the patient!**

9 Sacral Rhizolysis

Indication

- Pain of any origin, along the pathway of the first sacral nerve root, which cannot be controlled by other conservative measures.
- Pain of malignant origin along the pathways of the lower sacral nerve roots. It is very suitable for the pain which sometimes follows perineal resections for carcinoma of the rectum.

We do not recommend the use of this technique for lower sacral nerve root pain of benign origin, as there is a very real risk of impairing bladder function. We strongly advise that you do not use this technique for perineal pain of non-malignant origin, owing to the uncertain aetiology of the condition and the fact that many of these patients have a strong psychological component to their pain.

Warning!
Whereas in cervical, thoracic and lumbar rhizolysis one aims to destroy the dorsal root ganglion, the technique of sacral rhizolysis lesions the nerve itself. There must therefore be an increased possibility of deafferentation pain. We have not seen deafferentation pain as a result of gangliolysis, but it remains a theoretical possibility.

Clinical Picture

- S1 pain: the patient presents with pain along the distribution of the first sacral nerve. More often than not, the patient has had previous back surgery and the surgeons are not very keen to re-explore the back! The pain may have all or some of the characteristics of a neuralgia, i.e. burning pain with a shooting component, hypersensitivity, areas of patchy sensation, allodynia, paraesthesiae and dysaesthesiae. Conservative measures will have failed to help. A diagnostic nerve root injection (see page 85) would have given excellent pain relief.
- S2–S4 pain: pain along the lower sacral nerve roots, usually neuralgic in character and exhibiting the

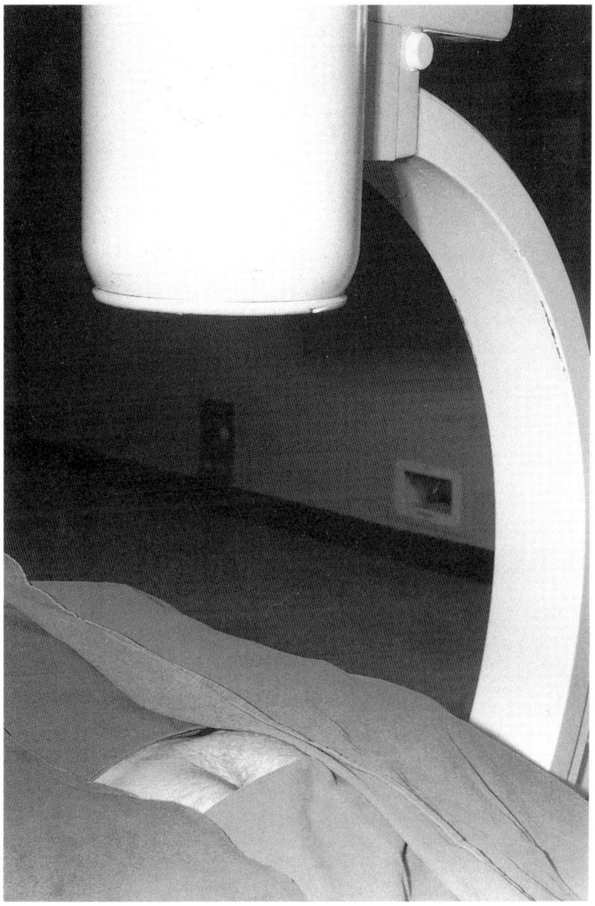

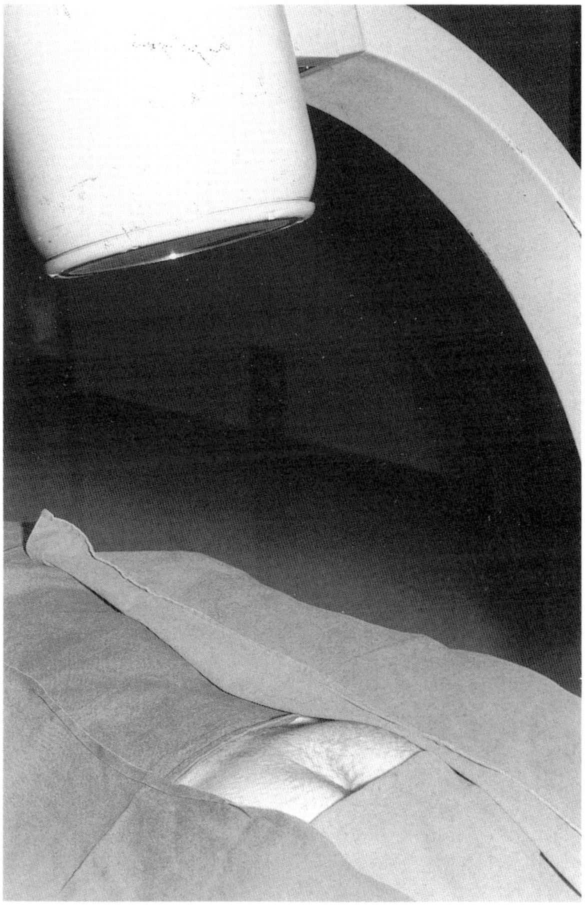

Figures 1a and 1b. Positioning of the X-ray tube to show the sacral foramina.

same features as described above. The pain is not controlled by conservative measures.

Technique

Sedation

Sedation is often necessary.

Position

The patient lies prone on a radiolucent table with a pillow under the abdomen in order to straighten the spine.

Landmarks

With the C-arm image intensifier positioned for screening in the anteroposterior plane, the level of the L5/S1 intervertebral foramen and the sacral foramina are identified; the image intensifier should now be positioned so that it is at exactly at right angles to the sacrum. (Figure 1a, b).

Clean and drape the area.

Procedure

There are three possible entry points for S1 rhizolysis, viz. directly through the S1 foramen; through the L5/S1 foramen; drilling through the dorsal aspect of the sacrum onto the S1 nerve.

Through the S1 Foramen

Mark the position of the first sacral foramen on the skin.

- Infiltrate the entry point with 0.5% Lignocaine with Adrenaline. Direct a 5 cm (2 in) needle towards the foramen and infiltrate; do not place any anaesthetic in the foramen itself.
- Insert the 100 mm Sluijter–Mehta cannula or pole needle through the entry point and direct it into the foramen under X-ray control; you will feel a definite 'give' as it enters the foramen (Figure 2a).
- Rotate the image intensifier so as to obtain a lateral view of the sacrum; the cannula/needle should be lying in the sacral canal. (Figure 2b). When you are satisfied with the position, if using a Sluijter–Mehta

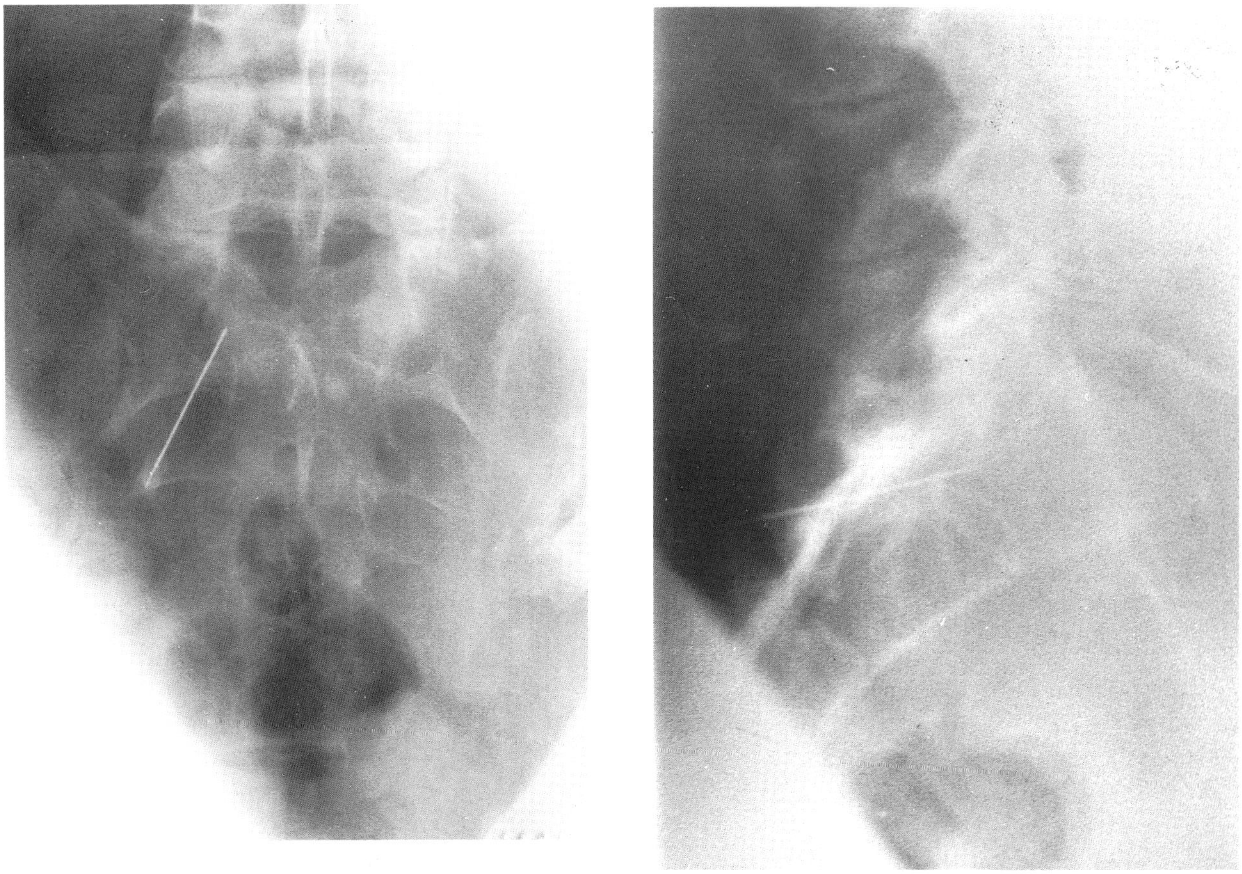

Figure 2.
a) PA X-ray, pole needle inserted through sacral foramen.
b) Lateral X-ray, pole needle inserted through sacral foramen.

cannula, remove the stylet and insert the thermocouple probe.

- With, the radiofrequency generator in stimulation mode, try to elicit pain/strong paraesthesiae in the area of the patient's pain at 0.5 V or less and at a frequency of 100 Hz. When you have achieved this, change the frequency to 2 Hz and ensure that you do not elicit motor contractions in the root distribution of that level at less than twice the voltage necessary to achieve sensory stimulation.
- When the needle is correctly positioned, if using a Sluijter–Mehta cannula, remove the thermocouple probe and inject 1 ml of 2% Lignocaine through it. Reinsert the thermocouple probe, wait for 1–2 min and then lesion at 70 °C. We recommend that you increase the current gently, allowing the temperature to increase, and when it reaches 65 °C, activate the timing switch, which is preset to 60 s; by the time this is done, the temperature will have reached 70 °C and should then be kept at that level.

  If using a pole needle, lesion at 22 V for 60 s (exact settings will vary with your machine).
- We strongly recommend that you lesion initially at a low temperature and repeat the procedure on several occasions at a slightly higher temperature in an effort to get a permanent effect. We would not recommend a lesioning temperature greater than 80 °C for fear of damaging the motor root.

If you are having difficulty locating the first sacral nerve, you may try injecting 0.2 ml aliquots of a water-soluble dye such as Omnipaque in an effort to demonstrate the nerve and thus give you a target at which to aim your needle.

Through the L5/S1 Foramen

Insert your needle through the L5/S1 intervertebral foramen, using the technique described on page 19, but angle the needle caudally and advance it so as to make contact with the S1 root (Figure 3). Proceed as above.

Drilling Through the Dorsal Aspect of the Sacrum

This technique may be used if you fail to locate the nerve

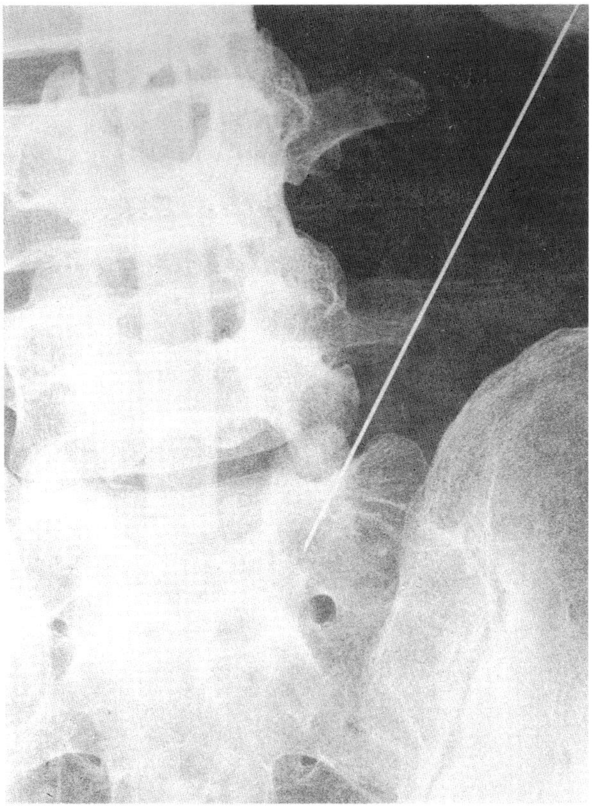

Figure 3. Pole needle inserted through the L5/S1 intervertebral foramen on to the S1 root.

root by either of the two techniques detailed above. The radiofrequency generator setting should be switched to 'thermistor'; a lead with bulldog clip should be autoclaved, plugged in and ready to use.

- Insert a 20/22 gauge 8.75 cm (3½ in) spinal needle into the first sacral foramen and inject water-soluble contrast medium as indicated above until you can demonstrate the first sacral nerve. When you have done this, mark the skin anywhere over the course of the nerve. Through this, using a hand-held orthopaedic drill and Kirschner wire, drill the back of the sacrum until you feel a 'give' as the wire penetrates the bone (Figure 4).
- Attach the bulldog clip at the end of the radiofrequency generator lead wire to the end of the Kirschner wire; together with the earth plate, you now have a complete circuit exactly similar to a pole needle (Figure 5).
- With the radiofrequency generator in 'stimulation' mode, advance and withdraw the Kirschner wire very gently until you succeed in obtaining sensory stimulation along the first sacral nerve. You are in effect using your Kirschner wire exactly like a pole needle. Test for motor stimulation as described above.
- When you are satisfied with the final position of the Kirschner wire, lesion as if with pole needle at 18–20 V for 1 min. With this technique, you do not have temperature monitoring, so there is a greater risk of damaging the motor root than with the first two techniques described above.

The S2–S4 nerve roots
If you wish to lesion these roots, insert your Sluijter–Mehta cannula/pole needle through the sacral hiatus and direct it laterally seeking sensory stimulation (Figure 6). T. Nash (personal communication) recommends the insertion of the Sluijter–Mehta cannula or pole needle through the sacral hiatus and the aspiration of 30 ml of cerebrospinal fluid in order to collapse the sac containing the nerve roots; these are then sought by stimulation at 100 Hz and 2 Hz in the usual way. This manouevre will cause 'post-spinal headache.'

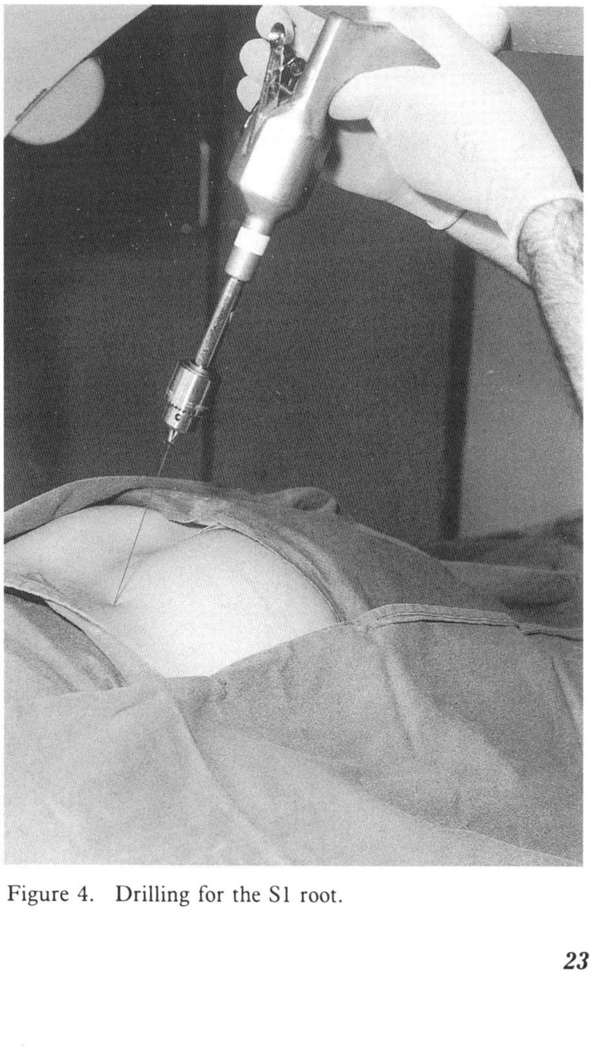

Figure 4. Drilling for the S1 root.

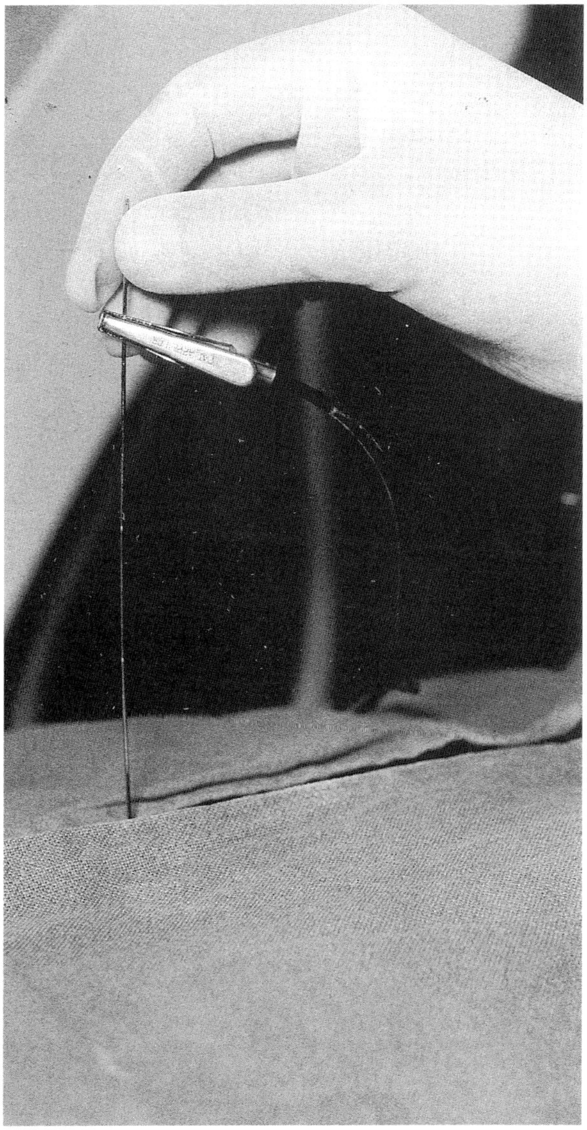

Figure 5. Stimulating through the Kirschner wire.

With this technique, you will **definitely** damage the motor fibres and the patient will become incontinent; this is why we recommend that this technique be used only for cancer pain. You may as well use a phenol caudal block!

After-care

It is prudent to keep the patient in hospital for at least 1 h after the lesioning. If cerebrospinal fluid aspiration has been carried out, the patient must be kept in hospital for a few days.

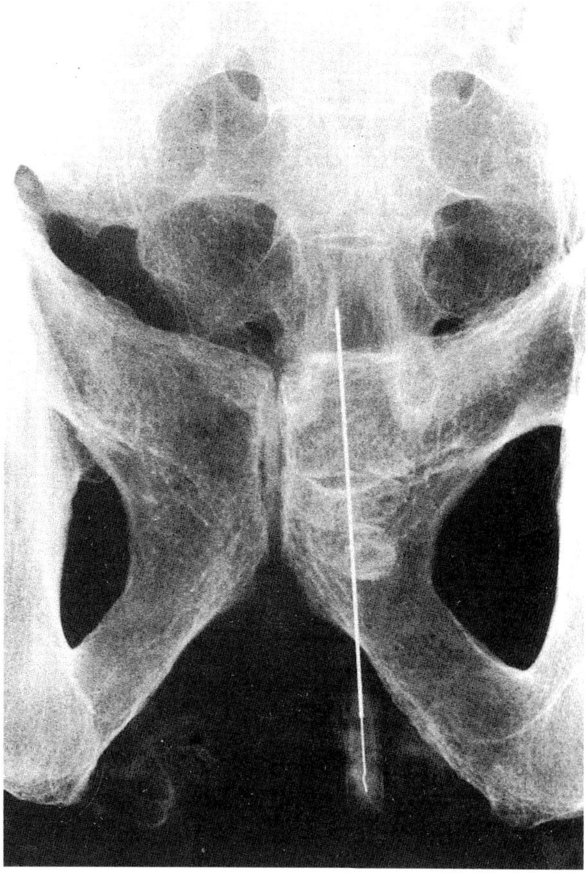

Figure 6. Pole needle inserted through sacral hiatus.

Complications

Immediate

- Aspiration of blood: resite needle.
- Motor nerve damage: this is unavoidable with this technique.

Late

General soreness in the area of the rhizolysis with or without protective muscle spasm. This may last for a few weeks, so **warn the patient!**

10 Trigeminal Thermocoagulation

Indication

Trigeminal neuralgia which cannot be kept under control by medication; the procedure should only be undertaken at the specific request of a neurologist.

Equipment

Our radiofrequency equipment of choice is the Radionics TIC kit (Figure 1a). This provides you with four 19 gauge hard Teflon insulated cannulae, each having a different length of non-insulated tip, viz. 2 mm, 5 mm, 7 mm and 10 mm. Into the cannula of choice (see below) fits a 22 gauge TM (temperature-monitored) electrode. The size of the lesion you produce will depend on the length of non-insulated tip you select. The fibres in the trigeminal ganglion are laminated (Figure 2); thus, the choice of cannula will depend upon whether you are planning to lesion the first, second or third divisions of the trigeminal nerve (V1, V2 or V3). You should use the 2 mm or 5 mm cannula for lesioning one division of the V ganglion, the 7 mm cannula for lesioning two divisions and the 10 mm cannula for lesioning all three divisions.

A very useful modification is the Type TEW kit (Figure 1b). Here you have a single Teflon-coated cannula through which you can insert a straight electrode or a curved-tip electrode. The straight electrode's bare tip can be pushed through the cannula's end by a variable amount, as indicated by millimetre markings on the cannula's hub; this allows an axially symmetric lesion to be made in the usual way. Should an enlarged lesion be needed in an off-axis direction to achieve desired analgesia, the straight electrode may be removed and a curved tip electrode inserted into the cannula; the curved flexible tip of the electrode extends beyond the end of the cannula also, by an amount indicated by the cannula's hub markings. The curved tip electrode may be of value if you are unable to obtain stimulation in the desired area, using the straight electrode.

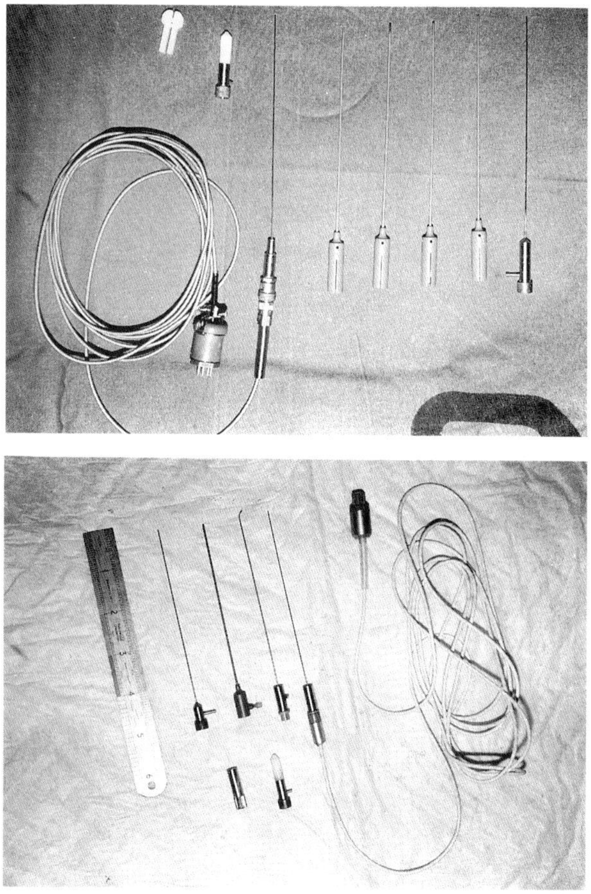

Figure 1.
a) Radionics TIC kit.
b) Radionics TEW kit.

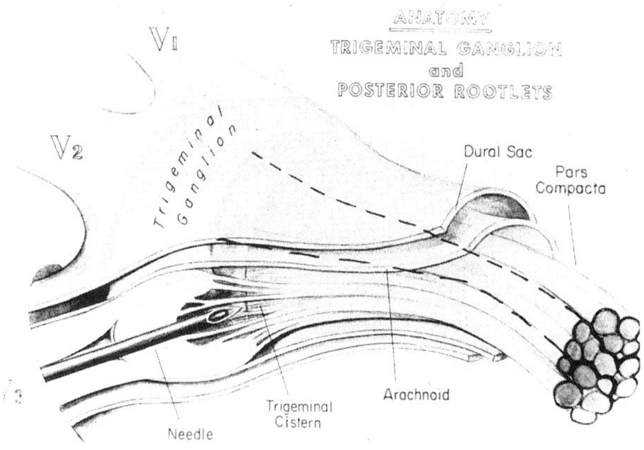

Figure 2.

Technique

Sedation

The operator will require the assistance of an anaesthetist; there is no room for the 'operator-anaesthetist' in this procedure. Trigeminal thermocoagulation carries the real risk of a hypertensive crisis which requires skilled therapy (5). In our experience, the combination of alfentanil and propofol anaesthesia offers excellent operating conditions.

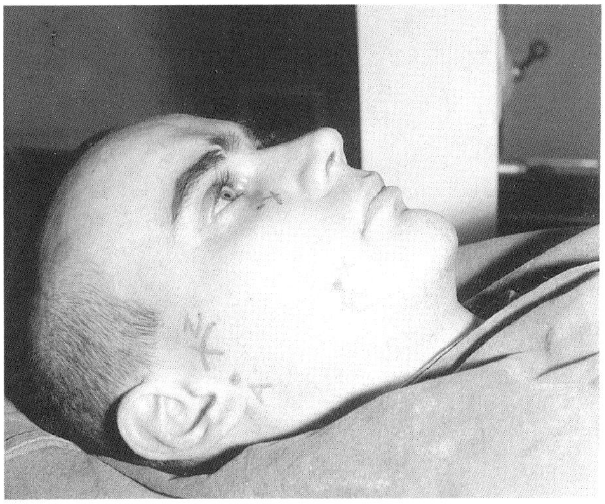

Figure 3. Landmarks for trigeminal thermocoagulation.

Position

The patient lies supine with his head on a radiolucent rest affording full access to the neck. The C-arm image intensifier is positioned for screening the anteroposterior plane.

Landmarks

The following points are mapped out on the patient's face (Figure 3).

Point X: about two fingers' breadth lateral to the angle of the mouth at the level of the 2nd upper molar tooth (the upper 6th).

> **Note**
> With regard to point X, the distance of 'two fingerbreadths' is a general guide. If you wish to lesion V2 and V3, we advise that you use this point, so that you can then place your electrode in the centre of the foremen; if you wish to lesion V3 alone, we recommend distance of one finger's breadth; so that you can then place the electrode in the lateral part of the foramen. For V1, the distance should be three fingers' breadth lateral to the angle of the mouth, so that the electrode will come to lie more medially in the foramen.

Point Y: in line with the pupil at the point of exit of the infraorbital vessels.

> **Note**
> A useful manouevre is to ask the patient to stare straight upwards at the ceiling; point Y will be in line with the centre of the pupil.

Point Z: mid-way along the zygomatic arch.

> **Note**
> With regard to Point Z, we find it useful to mark out the lower border of the zygomatic arch on the skin and then to bisect it and mark out the midpoint.

Point A: at the tubercle of the zygoma.

Procedure

- Take the radiofrequency cannula of your choice, into which is inserted a solid stylet. Introduce the cannula at point X, directing it towards point Y in the anteroposterior plane and towards point Z in the lateral plane (Figure 4). You may need to put your finger in the patient's mouth to guide the cannula and to prevent it from perforating the buccal mucosa. If the cannula does perforate the buccal mucosa, you will need to institute antibiotic cover. Advance the cannula gently until it hits bone.
- When the cannula hits bone, the beam of the image intensifier should now be rotated so that it follows the angle of the cannula; this should give you a good view of the foramen ovale (Figure 5).

> **Note**
> If your X-ray beam follows the line of the needle too perfectly, the head of the needle can occasionally obscure the view of the foramen ovale! (Figure 6).

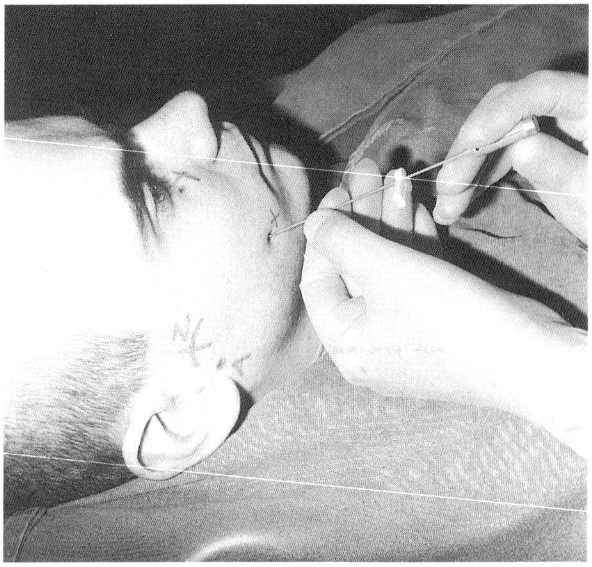

Figure 4. Inserting the radiofrequency cannula.

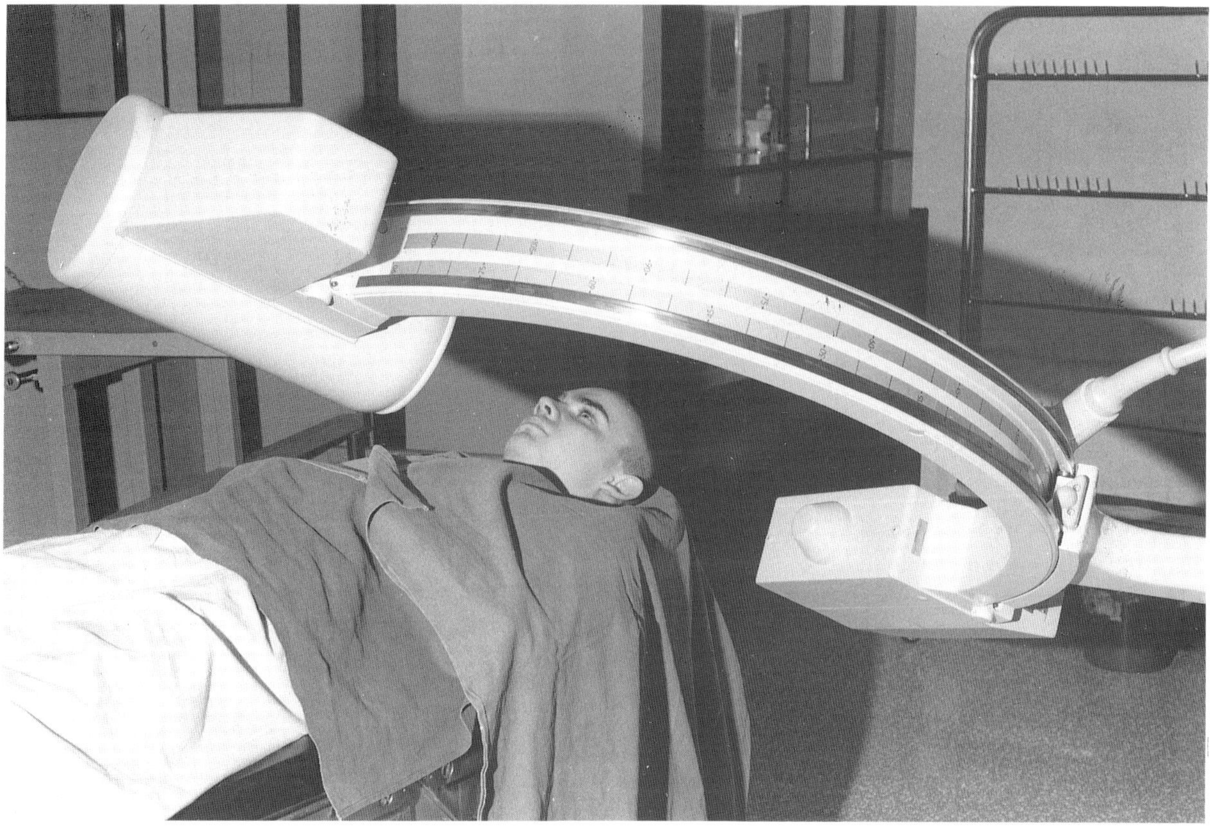

Figure 5. Position of the X-ray tube along the axis of the trigeminal cannula.

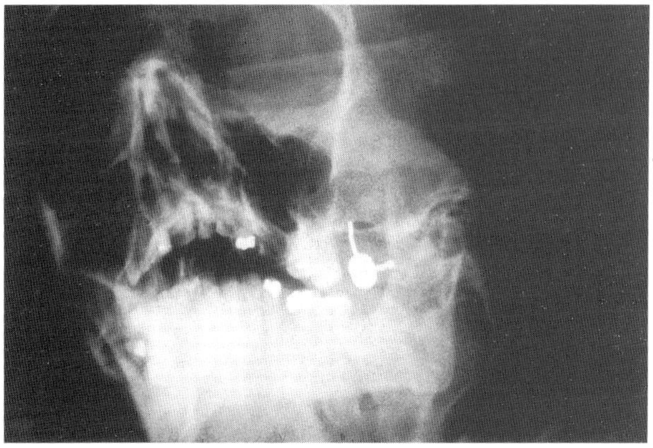

Figure 6. Cannula inserted through foramen ovale.

- (1) *If you can see the foramen ovale clearly.* Place a marker on the cannula 1.5 cm away from the skin; withdraw the cannula by about 5 cm and redirect it so that it is still pointing towards point Y in the anteroposterior plane but is now pointing towards point A in the lateral plane, i.e. it is now directed downward. Advance the cannula, which should enter the foramen ovale at the 1.5 cm marker you placed earlier. If you have an excellent view of the foramen ovale, you may decide to advance the cannula into the foramen under visual control, but do not advance it further than the 1.5 cm marker until you have obtained a lateral X-ray view (Figure 7).

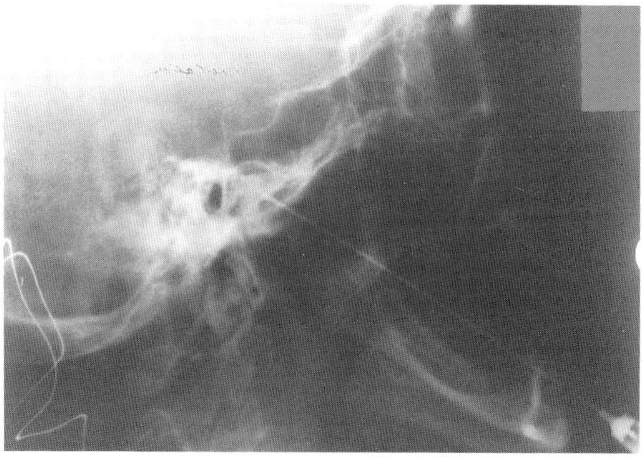

Figure 7. Lateral X-ray, cannula inserted through foramen ovale.

(2) *If you cannot see the foramen ovale clearly.* Place a marker on your cannula at skin level, then withdraw it by about 5 cm and redirect it so that it is still pointing towards point Y in the anteroposterior plane but is now pointing towards point A in the lateral plane. Now redirect the beam of the image intensifier along this new cannula angle, and this should enable you to visualise the foramen ovale. Advance the cannula into the foramen under visual control to the depth of the marker which you placed at skin level.

If, despite the above, you still cannot see the foramen ovale, you should take a static X-ray film, as this shows up detail much more clearly than does a monitor view. Redirect your cannula as necessary; you may need to take several static pictures is you redirect the cannula.

An alternative way of visualizing the foramen ovale is to align the X-ray tube as in Figure 5 at 45° to the vertical and then ask the patient to turn their head 35°–50° in the opposite direction.

Caution
Inserting the cannula into the foramen ovale will bring its tip in touch with the Gasserian ganglion. This can be very painful. Either precede the insertion by injecting a little Lignocaine or, since this can affect stimulation, ask the anaesthetist to inject intravenous propofol.

Note 1
Cerebrospinal fluid flow-back soon after entering the foramen ovale means that you have entered the trigeminal cistern and that your position is correct.

Caution!
It is always possible that you could inject local anaesthetic into the CSF. In this case you can get cardiovascular collapse which will need resuscitation; this can happen even with a negative aspiration test.

Note 2
If necessary, be prepared to advance a bit deeper through the foramen ovale for V1.

In the lateral X-ray view, the tip of your electrode should, if possible, come to lie just above the margin of the boney plate (Figure 7) so that you can lesion the preganglionic fibres and get a better effect.

Note 3
Should you produce masseteric, contractions on stimulation, this means that you are in touch with the motor root of the V; in this case advance your electrode tip deeper into the foramen and you should clear the motor root.

Note 4
Improper placement can take place as follows: superiorly into the inferior orbital fissure; posteromedially into the foramen lacerum (carotid artery); posteroinferiorly into the jugular foramen or carotid canal.

Note 5
Blood in large amounts means that you have punctured a large vessel. In this situation you should abort the procedure.

Stimulation

With the radiofrequency generator in 'stimulation' mode, try to elicit paraesthesiae/pain in the distribution of the division of the trigeminal nerve you are planning to lesion at less than 1 V and a frequency of 100 Hz. When you have done this, the anaesthetist should induce anaesthesia and you should then lesion at 60 °C for 60 s; the patient should then be allowed to recover from the anaesthetic and you should test for diminution of pinprick sensation. Remember that the area of the analgesia you induce will contract, so you must try to get a slightly larger area of diminution to pinprick sensation than the area of the patient's neuralgia. Carry on lesioning in 5 degree increments to a maximum of 80 °C for 60 C.

Note
Watch out for eye movements and facial contraction when the electrode has been inserted deeply through the foramen ovale; do not lesion in this situation, as you may damage the cavernous sinus!

Additional Comments

1. In some cases, e.g. fungating malignant tumours of the face, the patient may not be fit to undergo what is essentially a lengthy procedure. In these cases you may want to carry out an injection of absolute alcohol into the V ganglion. We recommend that you use 0.3 ml of absolute alcohol per division you wish to lesion. Precede this by the injection of an equal amount of 1% plain lignocaine and wait a full 5 min prior to injecting the alcohol, in order to ensure that you have not entered the cerebrospinal fluid.

2. A simple technique which may be of use in trigeminal neuralgia, if you do not have access to radiofrequency lesioning equipment, is to inject a mixture consisting of 2 ml of 1% plain lignocaine, 40 mg of dexamethasone and 7 ml of 0.9% saline. Use a 22 gauge spinal needle and ensure that you do not draw back any cerebrospinal fluid. If you do, the procedure has to be abandoned. The solution should be given in 2–3 ml aliquots about 5 min apart. Any beneficial effect may not be noticable for 48–72 h. The injection may be repeated as necessary every few months.

11 Percutaneous Cervical Cordotomy

Indication

Unilateral pain due to malignant disease, when the pain is not responsive to drug therapy or other less invasive methods of treatment. The patient's total co-operation is required for this procedure, as he will be expected to lie very still with his head resting on a clamp for about 1 h; in addition, he will have to guide you as to the site and area of any sensations/analgesia; you are advised to make it clear to the patient that you can stop the procedure at any time and come back another day, should he so desire.

This technique is safest when performed for pain originating in the lower half of the body. In the cervical cord the fibres involved in respiration lie close to the fibres from the upper limb region and furthest from the fibres from the lower limb region; hence, there is a greater risk of damaging respiratory fibres when trying to lesion upper limb fibres and less risk when trying to lesion lower limb fibres (Figure 1). A potentially

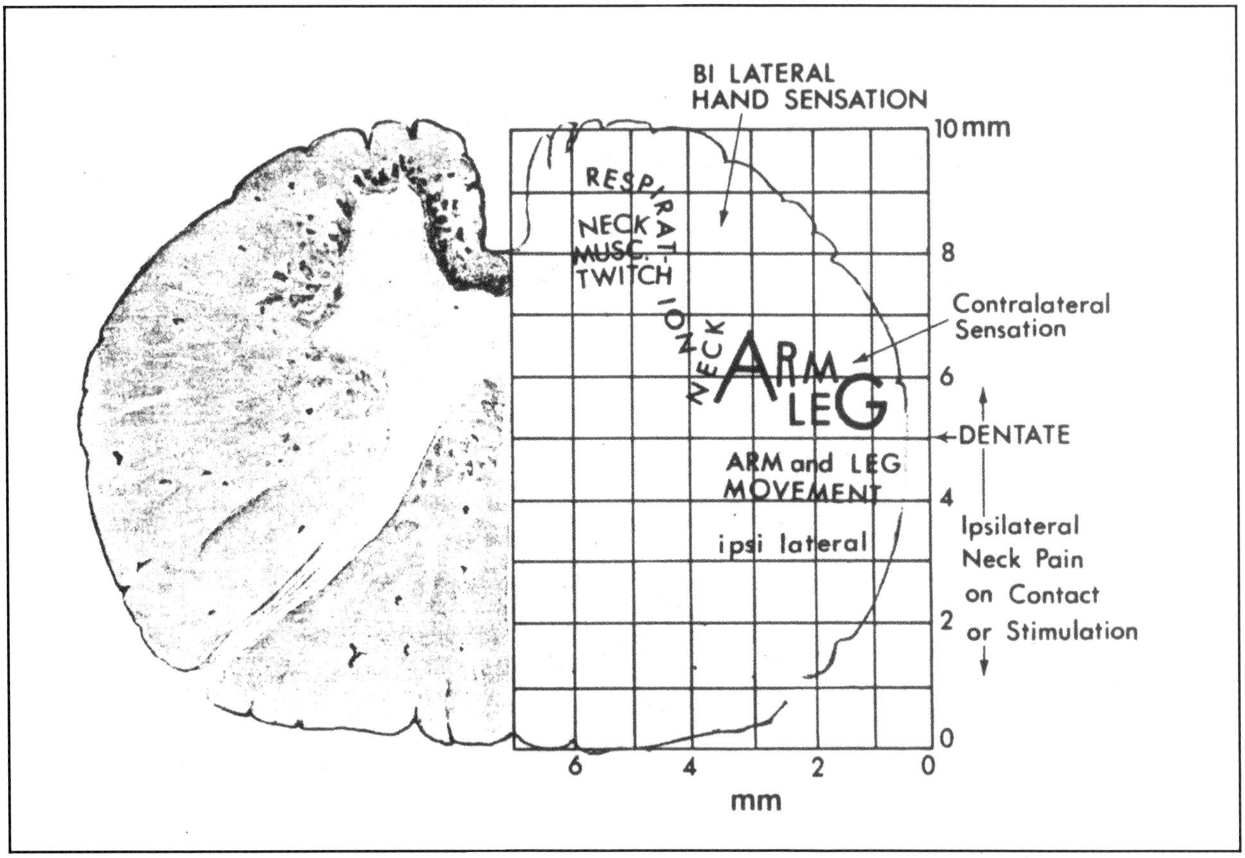

Figure 1. Anatomy of the cervical spinal cord.

fatal situation could arise if you have a patient with a Pancoast's tumour of the right upper lobe of the lung for whom you undertake a left-side cordotomy, possibly damaging the ipsilateral motor fibres; this would jeopardise the respiratory muscles on the non-diseased side, and since the respiratory function on the diseased side is already seriously impaired, the patient could very well die. Nevertheless, this technique remains the only effective form of treatment in some cases of Pancoast's tumour where the tumour has invaded the brachial plexus and is giving rise to severe pain.

Technique

Sedation

Mild sedation only, if necessary.

Position

The patient lies supine with his head an a ROSOMOFF neurosurgical head rest (Figure 2a, b); a practical point is that you should remove the top section of the operating theatre mattress, as otherwise the patient will be too high on the operating table to fit comfortably onto the head rest.

Landmarks

You will be working on the side opposite to the painful area; identify the mastoid process an this side and mark its apex; draw a mark one finger's breadth caudal to the tip of the mastoid; this will overlie your entry point (Figure 3).

Clean and drape the area.

> In view of the large cerebrospinal fluid leak that this procedure inevitably produces, we recommend antibiotic cover; select agents that are specific for Gram-positive organisms and also provide broad spectrum cover, e.g. Ampicillin combined with Flucloxacillin.

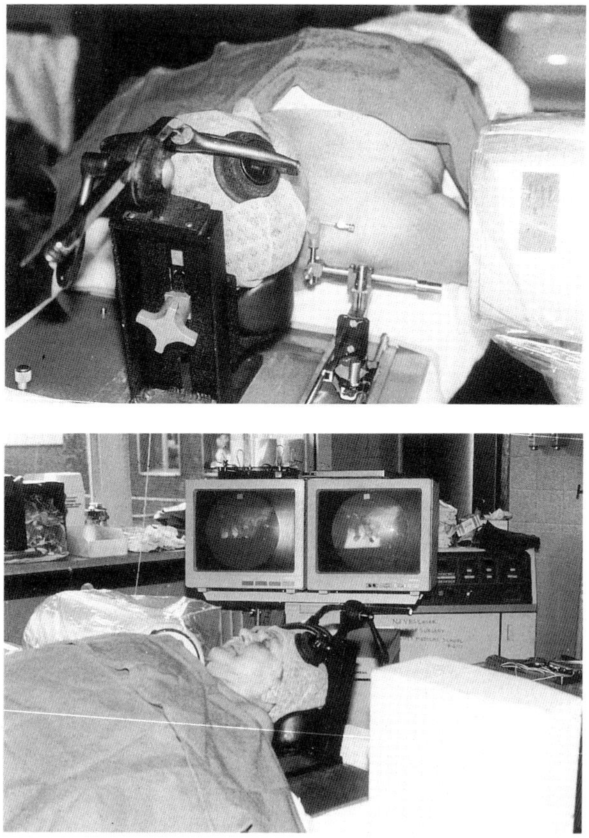

Figures 2a and 2b. Position of patient using ROSOMOFF head clamp.

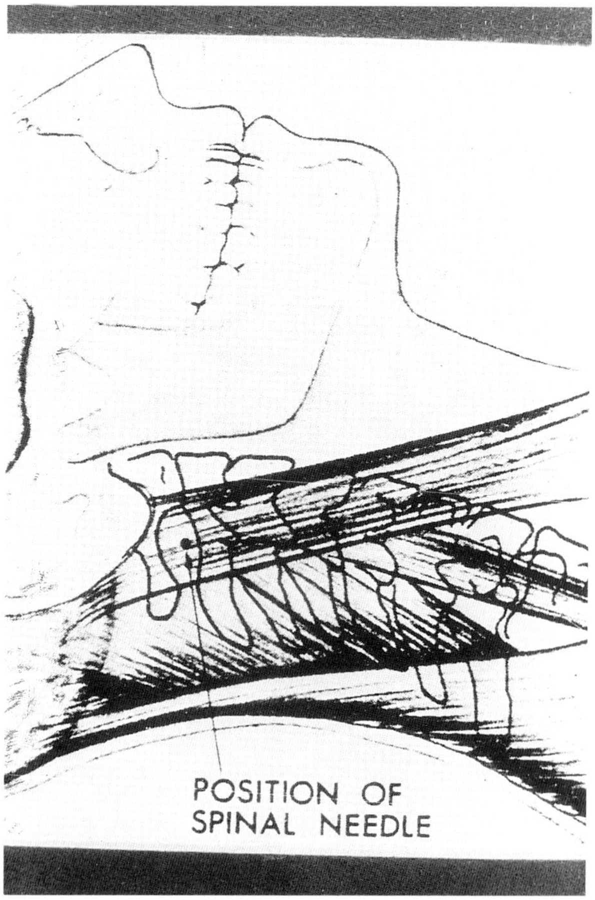

Figure 3. Entry point for percutaneous cervical cordotomy.

Procedure

- Infiltrate the entry point with 0.5% Lignocaine with Adrenaline.
- Your target area as seen on the image intensifier (lateral view) is the C1–C2 intervertebral space; this looks not unlike a house in shape, and your aim is to insert your needle just above the midpoint of the junction of the 'roof' and 'walls' of the 'house' (Figure 4). Adjust the length of the exposed tip of the cordotomy needle at this stage to between 3 mm and 7 mm, depending on the height-level of the required lesion and the size of the patient. The longer the exposed tip, the higher the eventual level of the lesion you produce.
- Insert the radiofrequency cordotomy needle through the entry point, directing it towards the zone described above. Attach it to the ROSOMOFF clamp (Figure 5a, b), then withdraw the stylet from the cordotomy needle and insert the thermocouple electrode. Note the impedance on the radiofrequency generator. Advance the needle gently centrally so that it enters the cerebrospinal fluid; the impedance will now drop to about 200 Ω.
- The next stage is to ascertain that your cordotomy needle is above the level of the dentate ligament. The importance of the dentate ligament is the fact that it lies dorsal to the lateral spinothalamic tract; thus, if your needle is situated ventral to the ligament, it should enter the tract (Figure 6).

  A useful way to visualise the dentate ligament is to prepare equal volumes of the radio-opaque dye Omnipaque in air. This can be done by drawing up 5 ml of dye in a 10 ml syringe and then adding 5 ml of air. Shake the mixture immediately between each injection. Inject 2 ml aliquots down your needle and the contrast provided by the air usually allows you to see the ligament (Figure 7). Injection through a remote entry point allows continuous screening during injection.

  If, upon injection, you are unable to see the ligament, this means that the dye has travelled cranially or caudally; in this case adjust the operating table tilt to compensate and the ligament should come into view.
- If you are satisfied that the needle is ventral (i.e. 'above' the dentate ligament), the next stage is to

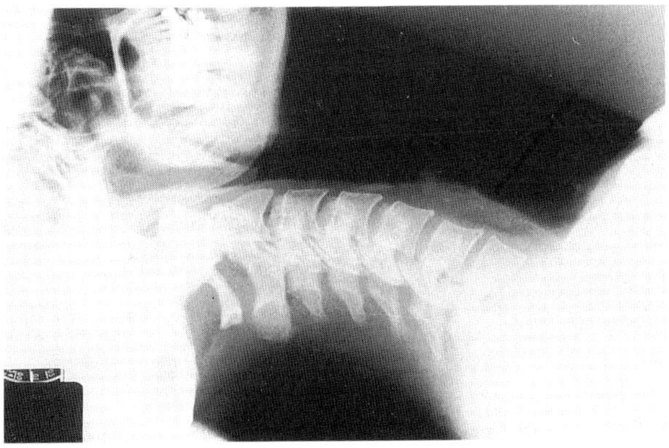

Figure 4. Lateral X-ray of the cervical spine.

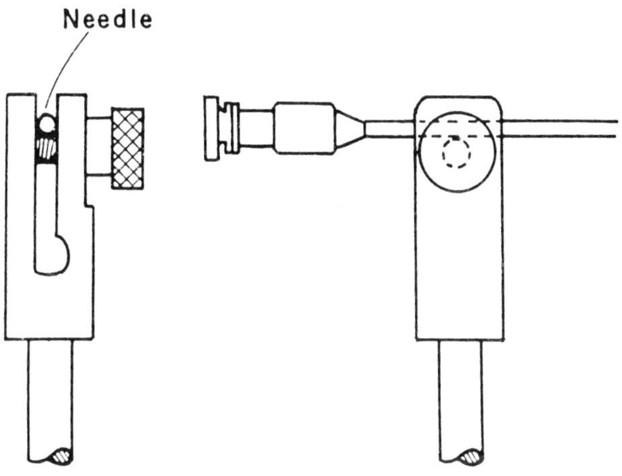

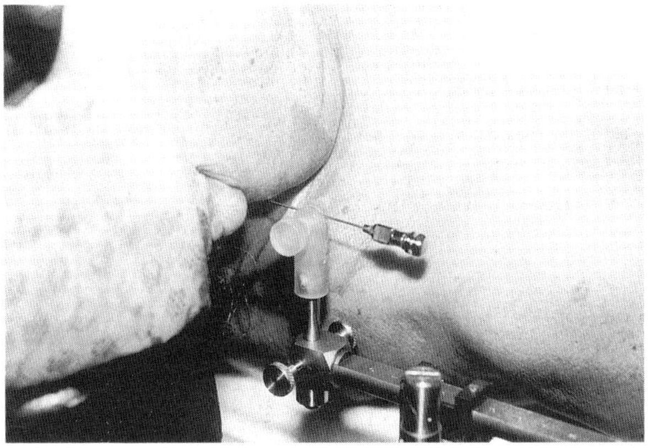

Figures 5a and 5b. Needle in clamp.

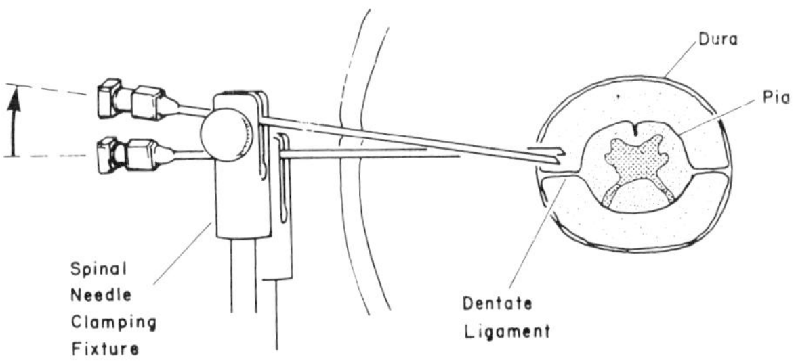

Figure 6. Needle in clamp showing entry into spinal cord.

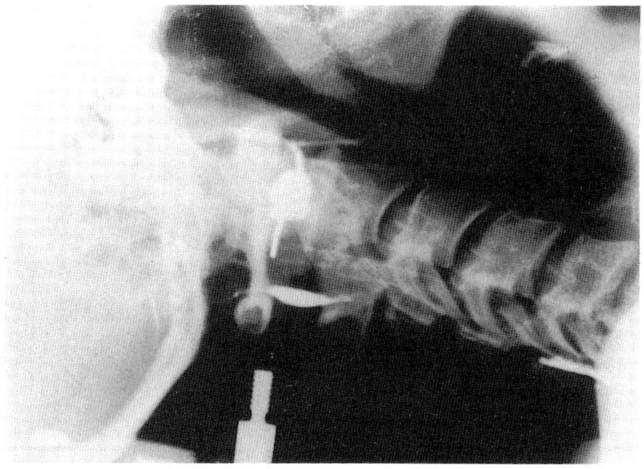

Figure 7. Cervical cord showing dented ligament after injection of air and contrast medium (Myodil).

enter the cord. As soon as you enter the substance of the spinal cord, the impedance on the radiofrequency generator should register about 700 Ω. The actual impedance measurements will vary from patient to patient: the important thing to remember is that when the electrode tip leaves the cerebrospinal fluid and enters the substance of the cord, the impedance level rises sharply. The cord may produce tough resistance to penetration: as you have already sized the exposed tip of the needle (see above), it is recommended that you enter the cord with a sharp, rapid movement. An overcautious attempt to enter the cord at this stage may cause the cord to 'bounce' away from the electrode tip. Figure 8 shows the correct position of the needle as viewed in the anteroposterior axis.

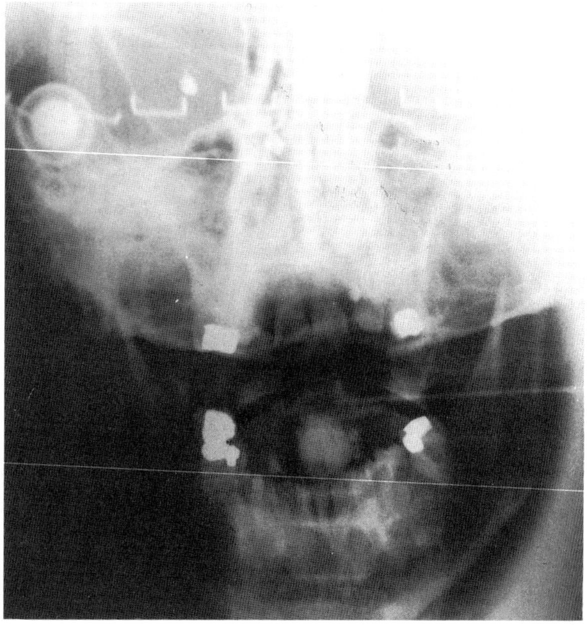

Figure 8. PA X-ray of needle in position for cordotomy.

- With the radiofrequency generator set in 'stimulation' mode, you should elicit sensory stimuli at less than 0.2 V on a frequency of 100 Hz; the sensations elicited can vary from pinprick to unusual sensations, e.g. 'a cold wind'. If stimulation is not elicited in the desired area, withdraw the thermocouple electrode from the cordotomy needle. The needle, firmly secured in its Rosomoff lamp, should be angled ventrally or dorsally. Reinsert the thermocouple electrode and stimulate as before. If, despite this manouevre, you still cannot elicit sensory stimuli in the desired area, you may have to pull the cordotomy needle back out of the cerebrospinal fluid and reinsert it at a different angle. The tip of the cordotomy needle should, in its final position, as viewed in the anteroposterior axis on the image intensifier, come to rest at the base of the odontoid peg in line with its lateral border (Figure 8).

- When you have elicited sensory stimuli in the target area, create a lesion, using the following scheme:

    1. 40 °C for 5 s. Test for loss of appreciation of pinprick sensation in the target zone. Repeat the lesioning process for 10 s, 20 s, 40 s and 60 s as necessary, repeating the testing after each lesioning. Remember that the area of analgesia will shrink, so you should aim to get sensory changes over 1–2 segments each side of the painful target zone.

    2. If the temperature of 40 °C is not high enough to achieve the desired area of analgesia, you will have to lesion at temperatures of 45 °C, 50 °C, 60 °C, and 70 °C as necessary; before going on to a higher temperature you should lesion at the lower temperature for the time sequences mentioned above.

    When the desired area of analgesia has been obtained, repeat the lesioning at that particular temperature twice more in order to consolidate the lesion. After each lesioning you should test for sensory changes as already seen, and also ask the patient to lift his or her arm and leg and test for motor power on the same side as the side where you have inserted your needle, in order to make certain that you are not damaging any motor fibres.

We appreciate that the technique as described by us is somewhat tedious, but it is certainly very safe.

Caution!
Only in about 90% of patients do the pain fibres cross over by the C2 level; in about 10% of patients this does not happen. If good sensory stimulation cannot be obtained, you may have to abort the procedure.

After-care

The main problem is the cerebrospinal fluid leak. In order to minimise headache, vomiting and confusion, keep the patient in bed for a couple of days and give a high fluid intake – if necessary, by intravenous infusion.

Complications

- Meningitis: the risk of this should be minimised by using antibiotic cover.
- Mild motor weakness often appears on the same side as the lesion; this usually wears off after a few and is presumably caused by oedema. It can persist, however.
- Respiratory problems in high cordotomies – see above.
- Sympathetic fibres can be lesioned, leading to complications such as bradycardia (6)

(1) Ford, D.J. *et al.* (1984) Comparison of insulated and uninsulated needles for locating peripheral nerves with a peripheral nerve stimulator. *Anesth. Analg.*, **63**, 925–928.

(2) Hamann, W. & Hall, S. (1992) Acute effect and recovery of primary afferent nerve fibres after graded radiofrequency lesions in anaesthetized rats. *Brit. J. Anaes.*, **68**, 443P.

(3) Gauci C.A., Thorp, S. and Kidd, A.G. (1992) *Journal of the Pain Society of G.B. & I.* **10**, 1, 33–34.

(4) Bogduk, N., Macintosh, J. and Marsland, A. (1987) Technical limitations to the efficiency of radio frequency neurotomy for spinal pain. *Neurosurgery*, **20**(1), 529–535.

(5) Swerdlow B., Shuer L. and Zelcer, J. (1988) Coronary Vasospasm during percutaneous trigeminal rhizotomy. *Anaesthesia*, **43**, 861–863.

(6) Foster J.M.G., Cashman, J.N. and Jones R.M. (1984) Vasomotor disturbance at unilateral cordotomy *Anaesthesia and Intensive Care*, **12**, 376–7.

2

Cryotherapy

1 General Introduction

The cryoprobe consists essentially of two tubes fitted within one another; through the inner tube, which tapers to a fine nozzle, gas (N_2O or CO_2) is forced at high pressure. As it rushes out of the nozzle, the gas expands into the space between the two tubes and this expansion causes a rapid drop in the temperature of the cryoprobe tip by the Joule–Thompson effect. The expanded gas is vented out of the probe to atmosphere via a scavenging system in the outer tube (Figure 1).

The temperature drop causes the cryoprobe tip to extract heat from the tissues, with which it is in contact. An ice ball forms at the probe tip and the freezing process (to about −70 °C) causes degeneration at the axonal and myelin sheath but leaves the connective tissue element unaffected. Thus, the selected nerve is only temporarily inactivated, as regeneration will occur. This accounts for the inherent safety of the technique.

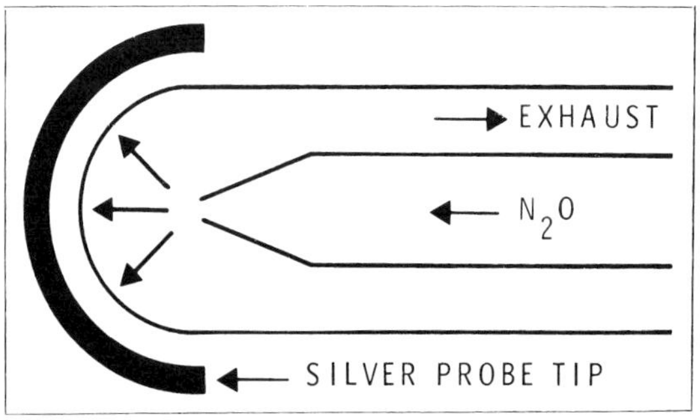

Figure 1. Structure of the cryoprobe.

The probe most commonly used in the UK is the Spembly cryoprobe, which comes in various sizes (Figure 2), supplied together with the Lloyd Neurostat apparatus (Figure 3). In this probe warm incoming gas insulates the patient's skin from cold exhaust gas; thus, cold damage only occurs where you want it to, i.e. at the probe tip.

A nerve stimulator is incorporated into the tip of the cryoprobe, and this may be used to help you localise the nerve you wish to lesion. In practise, however, the size of the cryoprobe precludes its insertion without a goodly preliminary dose of local anaesthetic, and this tends to interfere with the localisation process.

The cryoprobe can be used either with carbon dioxide or with nitrous oxide: the manufacturers will supply cylinder yokes for both gases or for either of them. The advantage of nitrous oxide is that it can make the cryoprobe tip reach extremely low temperatures, −70 °C, as opposed to −60 °C for carbon dioxide. The disadvantage of nitrous oxide is that its cylinders are very susceptible to changes in ambient temperature: if you do not allow the cylinder of nitrous oxide to equilibrate for 2–3 days in the operating room before use, you may get a rush of liquid nitrous oxide when you open the cylinder. This may clog the cryoprobe, especially the smaller 1H3 model. It is also a good idea to open the cylinder first before attaching it to the apparatus, in order to blast off any dirt or grit, as this can also clog the filters and the cryoprobe.

Use of the Lloyd Neurostat

The apparatus should be installed according to the manufacturer's recommendation; it is essential that the gas cylinder be kept upright or at an angle of 45 degrees.

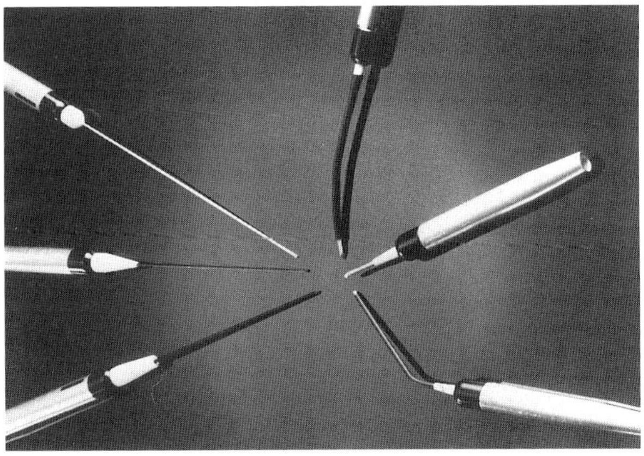

Figure 2. The Spembly range of cryoprobes.

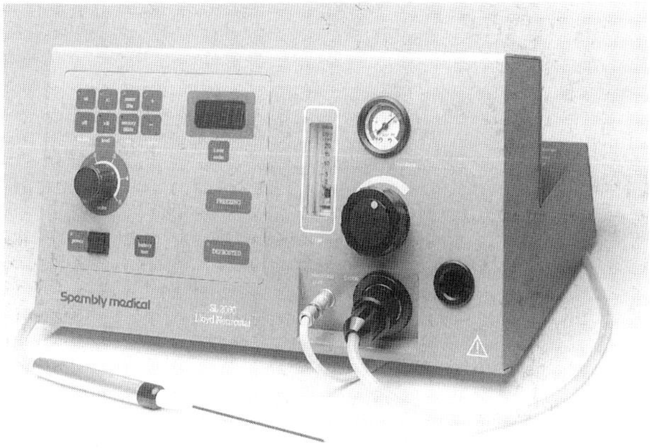

Figure 3. The Lloyd neurostat.

Purging the System

Before using the apparatus, you must ensure that all air and water vapour have been expelled from the cryoprobe.

A. Select the cryoprobe of your choice, remove the sterilising cap from the plug end and then plug in to the 'probe' orifice on the neurostat.

B. Place the cryoprobe tip in a receptacle containing sterile saline solution.

C. Turn the pressure control setting on the front panel so that the needle on the pressure gauge enters the 'purge' zone (colour-coded orange on the dial). At this setting the flow gauge should read just less than 2 l/min; keep this up for 1–2 min and then turn the pressure control setting to 'off'.

You are now ready to use the apparatus.

Nerve Location

A. Confirm that the electrode plate has been fastened to the patient and is linked to the neurostat.

B. Take an intravenous cannula of suitable length, such as a 140 mm 'Medicut' or an 'Angiocath', and insert the cryoprobe through it, noting the area of exposed cryoprobe protruding through the tip of the cannula when the cryoprobe is pushed fully through it. The size of cannula you use will depend upon the size of the cryoprobe. Thus, the 1H3 cryoprobe needs a 14G cannula and the 2T10 cryoprobe needs a 12G cannula.

C. Use the intravenous cannula to pierce the skin overlying the nerve you wish to locate and push the cryoprobe through it to make approximate contact at the apposite depth.

D. Set the 'volt control' to zero, switch the 'locator' on and set the 'level' switch to '$\times$ 10'. Choose 'sensory' or 'motor' setting and 'positive' or 'negative' polarity for the probe; this latter determines the direction of the current, since stimulation is always greatest at the negative pole.

E. To locate the nerve, increase the 'volts' until the desired response is obtained, i.e. tingling or muscle contraction (depending, of course, upon whether you are seeking a sensory or a motor nerve). This brings you into approximate contact with the nerve in question. By gentle repositioning of the cryoprobe, try to obtain the desired effect at the lowest possible voltage setting. Then reset the 'level' to '$\times$ 1' and repeat the procedure. When you have completed this manoeuvre, you are ready to freeze the nerve. Since freezing may be painful, we suggest that you pull the cryoprobe out of the cannula and then inject 2 ml of 2% lignocaine; reinsert the cryoprobe through the cannula and freeze.

Freezing

A. Choose whether to have the 'timer audio' switch on or off; this sends out a single signal beep every 30 s and multiple beeps every whole minute.

B. Place your foot firmly on the 'foot switch'.

C. Slowly increase the pressure control, observing both the 'flow gauge' and the 'pressure gauge'; continue to increase the pressure until the needle on the 'pressure gauge' enters the 'freeze zone'; colour-coded green on the gauge. This setting should correspond to a flow of 9 l/min with the 1H3 probe and to 12 l/min with the 2T10 probe. As soon as the needle enters the 'freeze zone,' remove your foot off the 'foot switch'; the 'freezing' light will come on and the timer will start to operate. We recommend that you carry out two 'freezes', each of two min duration and separated by a defrost period of 1 min. This routine should be followed at all the sites you wish to lesion.

D. For an 'intra-operative defrost', put your foot back on the 'foot switch' and the 'defrost' light will come on after a short interval; only now should you relocate the cryoprobe if you want to. Remember that moving the cryoprobe while the tip is still frozen is dangerous, as it can tear the tissues. So long as your foot is on the 'foot switch', refreezing will not take place; if you wish, you can obtain a 'defrost' by actually decreasing the pressure on the machine, but this is not necessary until you have finished all your lesionings. To reactivate the cryoprobe, simply remove your foot off the 'foot switch': the freezing cycle will start once more.

E. When finished; Turn the pressure on the machine down, and when defrosting has occurred, extract the cryoprobe and the cannula. You must now vent the system by setting the pressure control to its maximum setting, with the cryoprobe still attached to the neurostat but held in midair; then switch 'power' off.

2 Sacral Nerve Roots

Indications

Pain originating from the lower sacral nerve roots and which has not responded to more conservative measures. It is useful for pain of benign origin where you do not want to run the risk of permanently impairing bladder function.

It is especially useful for the perineal pain which sometimes occurs about six weeks following abdominoperineal resections for carcinoma of the rectum.

General Technique

Sedation

Not usually necessary.

Position

The patient lies prone on a radiolucent table with a pillow under the abdomen in order to straighten the spine.

Landmarks

With the C-arm image intensifier positioned for screening in the anteroposterior plane (Figure 1a) the sacral hiatus

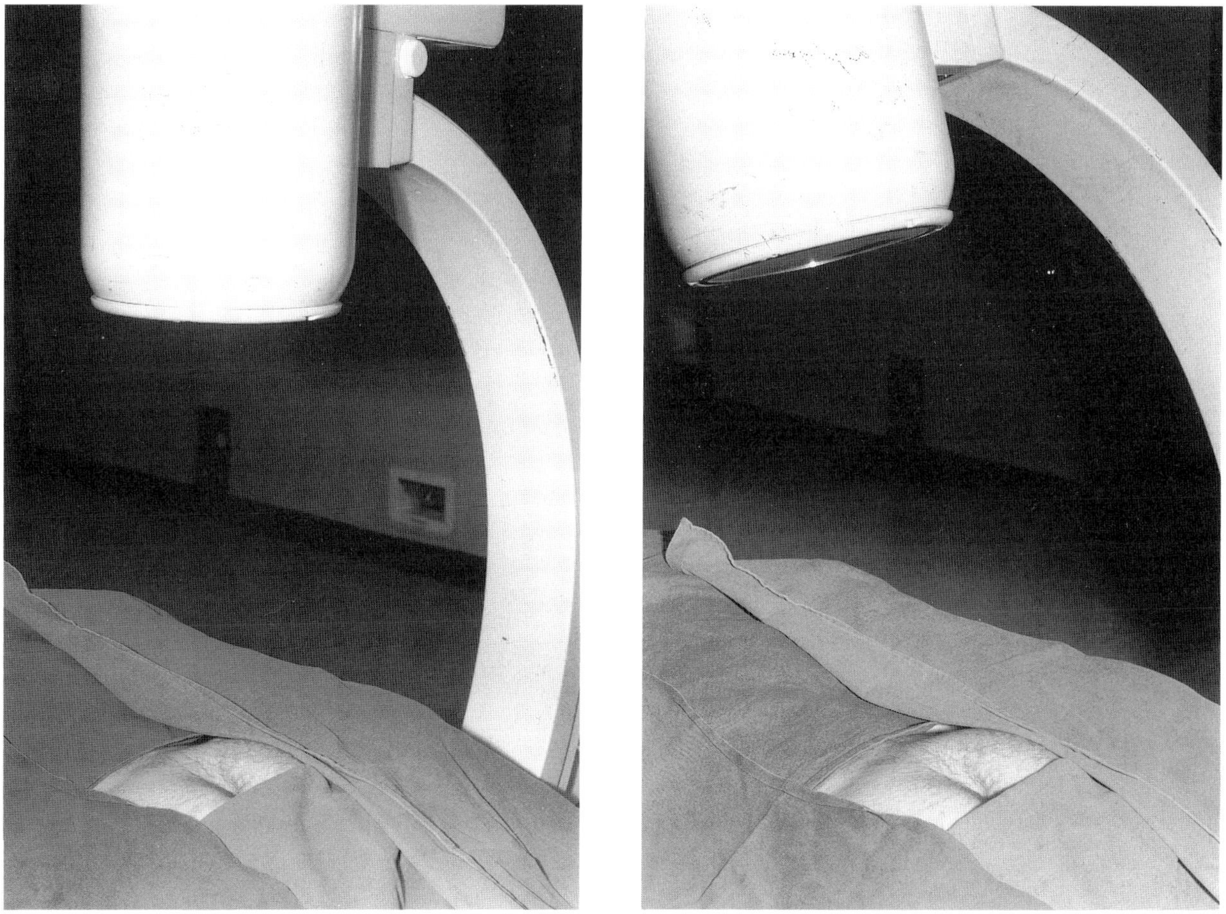

Figure 1
a) Position of the C-Arm for screening in the anteroposterior plane.
b) Position of the C-Arm at right angles to the plane of the sacrum.

is identified the image intensifier should now be positioned so that it is at exactly at right angles to the sacrum (Figure 1b).

Clean and drape the area.

Procedure

Insert an appropriate intravenous cannula through the sacral hiatus, directing it to one side of the sacrum, and through it introduce the cryoprobe (Figure 2). Rotate the image intensifier to screen the sacrum in the lateral plane; this will enable you to visualise the cryoprobe and to keep it below the second sacral segment (Figure 3). By avoiding lesions at the S2 level you should preserve bladder function. Produce freeze lesions at S3, S4 and S5 by the technique described above. When you have completed your work on that side, redirect your cannula and cryoprobe to the other side and repeat the procedure.

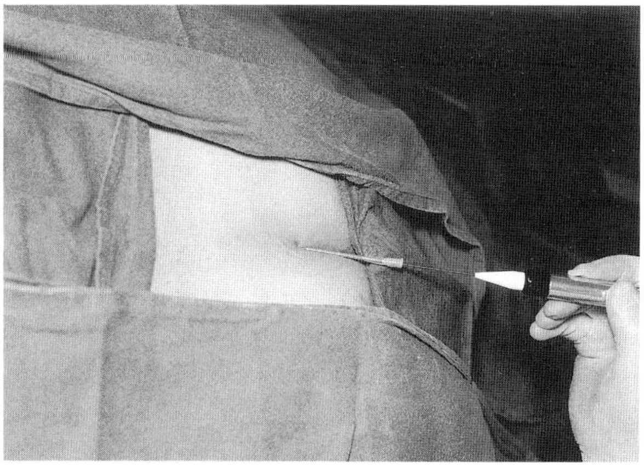

Figure 2. Introduction of the cryoprobe through an intravenous cannula into the sacral hiatus.

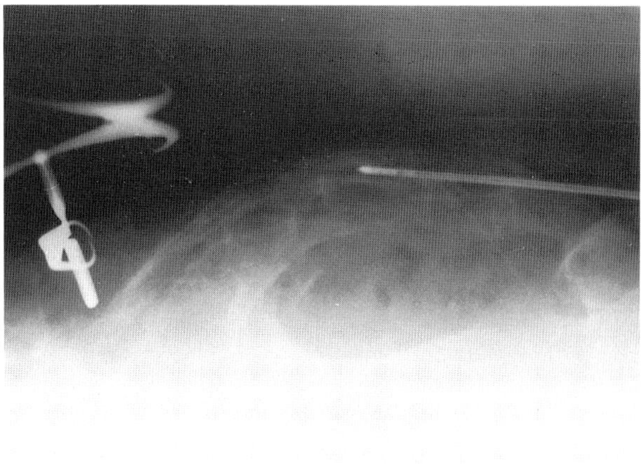

Figure 3. Lateral X-ray of the cryoprobe in the sacral canal.

3 Coccygeal Nerves

Indication

Coccydynia.

General Technique

Insert an appropriate intravenous cannula over the back of the coccyx; through it insert the cryoprobe and freeze at several levels, first on one side and then on the other.

4 Paravertebral Space

It is best to use image intensifier screening for cryotherapy work to the paravertebral space, as owing to the large size of the intravenous cannula and cryoprobe, there is a very real risk of pneumothorax.

5 Other Areas

The cryoprobe can be used to block the following areas quite successfully.

- Intercostal nerves.
- Suprascapular nerves – for some types of painful shoulder.
- Ilioinguinal nerves.
- Genitofemoral nerves – especially useful for the neuralgia which sometimes appears following a chemical sympathectomy.
- Occipital nerves.
- It can be used instead of the radiofrequency system for treating facet joint pain. For techniques of locating the nerves to the facet joints, see pages 93 *et seq*.
- It can be used to numb the sacroiliac joint. In these cases carry out freezing lesions at the superior, middle and inferior poles of the joint on both sides; it may be advisable, in addition, to carry out additional freezes at intermediate points along the joint line.
- It is very useful for dealing with neuromas, painful scars and 'trigger zones'.
- The pituitary gland can be lesioned by cryotherapy (see page 58), as can the trigeminal ganglion (see page 25).

3

Chemical Neurolysis

1 Intrathecal Neurolysis

Indication

Pain of malignant origin arising at any level and which cannot be adequately controlled by drugs or other more conservative therapy.

Preoperative Assessment

1. Patients for whom this technique is considered appropriate are often very ill, so they need to be handled with due care and attention.

2. It is always a good idea to carry out an initial subarachnoid block with heavy Bupivacaine; this will allow you to assess the level of blockade and the degree of pain relief. It will give you a good idea of the volume of solution needed in order to obtain the desired permanent block. A temporary block also shows the patient the degree of numbness he or she can expect to accompany the pain relief and will allow him or her to decide on the acceptability or otherwise of a permanent block. The problem caused by the numbness will vary from patient to patient. If the patient is bedridden, then numbness is not all that much of a problem, but if the patient is ambulant, intense numbness in a leg may adversely affect balance and gait, and may well prove unacceptable.

Be sure to examine the patient's fundi for the presence of papilloedema, especially if you are considering using phenol in glycerin, since you will need a large (19 gauge) needle to inject the chemical; the presence of raised intracranial pressure may well make it inadvisable to proceed.

3. With sacral nerve root ablation, bowel and bladder function must be taken into consideration, as they will always be at risk. If the patient is already incontinent, then no particular problems will be created by ablating the sacral roots. If the patient is continent, you will need to balance the pros and cons very carefully, as sphincter function damage may be unavoidable.

Technique

Sedation

Sedation is important, as you will need the patient to remain static for about 30 min after injection of the neurolytic agent into the cerebrospinal fluid. You will need a colleague to administer the sedation. We recommend that you leave the patient on all his or her medication and that a mixture of midazolam/alfentanil be used for sedation when necessary.

Position

Position will depend upon the root value of the pain to be treated; if a large area needs to be covered, you may decide to carry out the block at the uppermost/lowermost root level and to tilt the patient into Trendelenburg or anti-Trendelenburg, depending upon whether you are using a hypobaric (e.g. absolute alcohol) or a hyperbaric (e.g. phenol in glycerin) neurolytic agent. Figure 1 shows the spinal nerves is they exit the vertebral column; as can be seen from this figure, the root value coincides with the vertebral column segment.

Thoracic/Lumbar Blockade

The position will depend on several factors:

- If you are using phenol in glycerin, the patient must

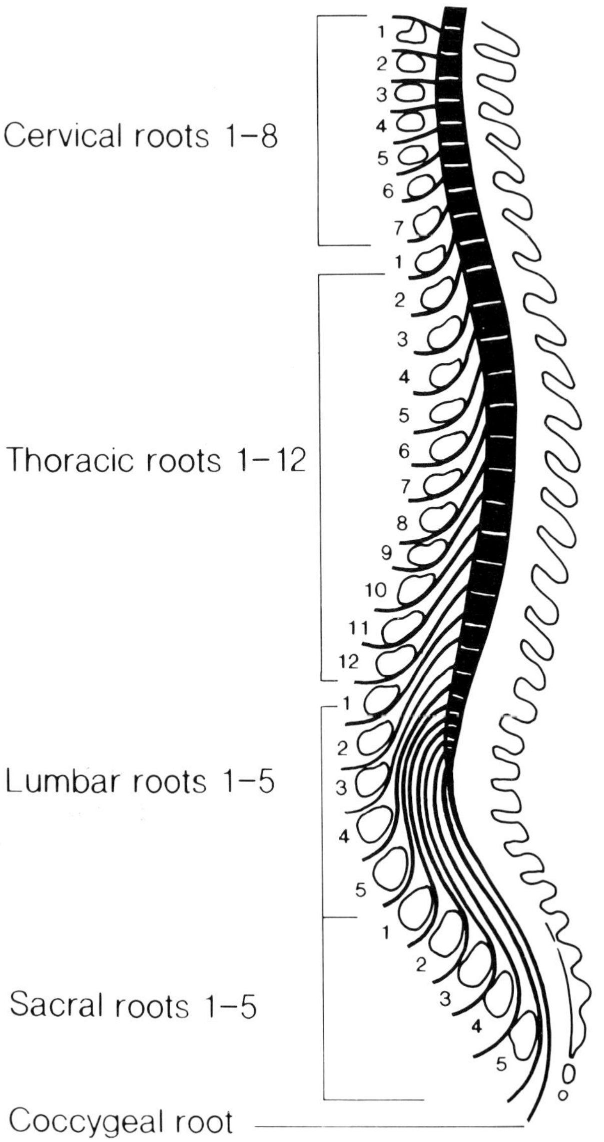

Figure 1. The spinal nerve roots.

be positioned with the painful side downward; if you are using absolute alcohol, the reverse is the case.

- If you are using phenol in glycerin, the patient must be titled at an angle of 45 degrees towards you when you inject the agent (Figure 2a); if you are using absolute alcohol, then the patient must be tilted at an angle of 45 degrees away from you on injection (Figure 2b).
- In *males*, the shoulder girdle is wider than the hip girdle; thus, when the patient lies in the lateral position, hyperbaric material deposited in the cerebrospinal fluid tends to drift caudally (Figure 3a). In *females*, the hip girdle is wider than the shoulder girdle, so in the lateral position hyperbaric material deposited in the cerebrospinal fluid tends to drift cranially (Figure 3b).

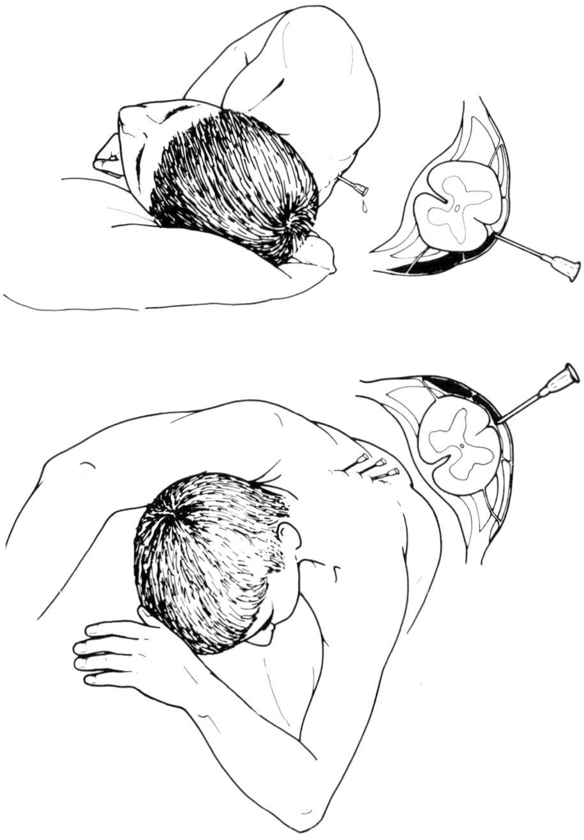

Figure 2.
a) Patient positioned for intrathecal injection of Phenol and Glycerin.
b) Patient positioned for injection of intrathecal alcohol.

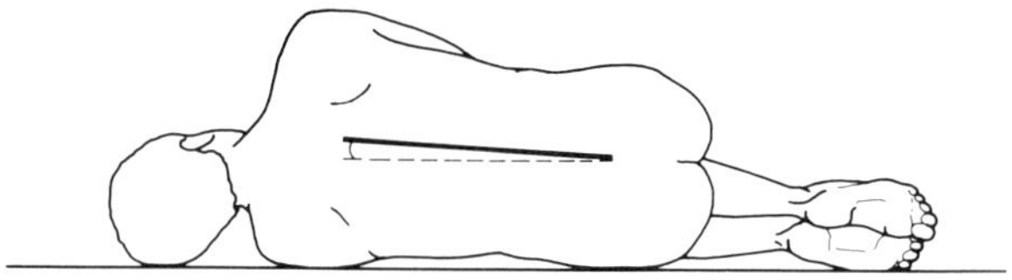

Figure 3a. The lateral position in the male patient.

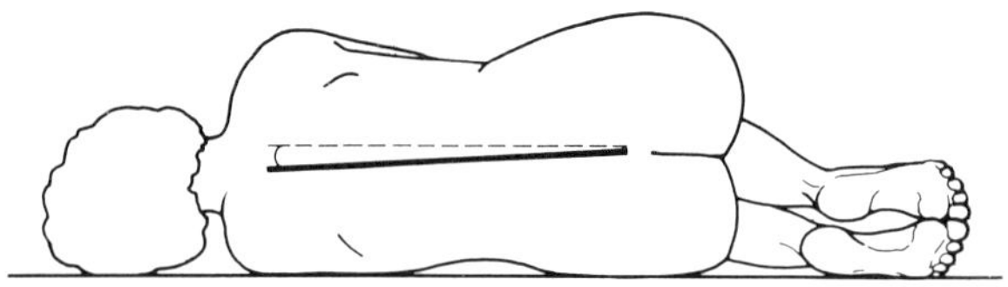

Figure 3b. The lateral position in the female patient.

When you first place the patient in the lateral position, use a skin marker to draw a line over the spinous processes from shoulder area to hip area; this allows you to assess the slope of the spine. You should then use pillows either to neutralise the slope of the spine so as to bring it into a straight line or to tilt the spine upwards or downwards, depending on where and what you want to inject (Figure 4a, b).

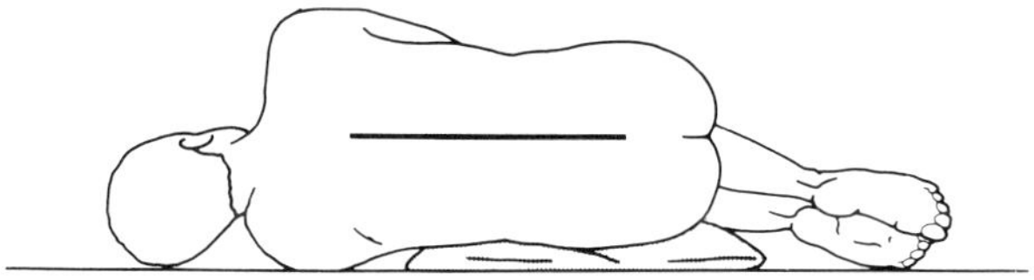

Figure 4a. The lateral position in the male patient corrected prior to injection.

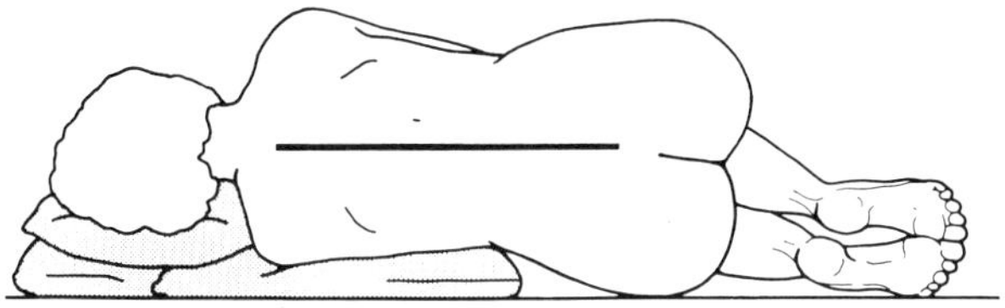

Figure 4b. The lateral position in the female patient corrected prior to injection.

When the slope has been dealt with, tilt the patient towards or away from you at an angle of 45 degrees on the criteria detailed above; this indicates the final position the patient must be in. Of course, it is almost impossible to carry out a lumbar puncture in this positions; thus, an assistant supports the patient in the lateral position for you to introduce your needle into the subarachnoid space.

When you have introduced the needle into the subarachnoid space, the patient is then tilted into the optimum position for the spread of the neurolytic agent, which should then be injected. Once you have done this, you should rapidly but smoothly extract the needle and spray the back with an antiseptic spray.

The patient should then be left in this position for 30 min, stabilised with an adequate number of pillows and suitably sedated so that there is no movement.

In some situations this sequence of events may not be possible, so in these cases inject the neurolytic agent with the patient in the lateral position, rapidly but smoothly extract the needle, spray the back with an antiseptic solution and then, also rapidly, place the patient in the position of optimum tilt.

Agents

- 5% Phenol in glycerin is a very commonly used agent; you can use the 10% solution at a second session if the 5% solution proves to be inadequate. If you are planning to ablate motor roots – e.g. to deal with uncontrollable pain from muscle spasm in disseminated sclerosis – you can use the 10% solution straight away. In these cases note that you should reverse the tilt employed for the sensory roots as described above; thus, if employing phenol in glycerin (hyperbaric), tilt the patient at angle of 45 degrees away from you. Phenol in glycerin is very viscous, so you will need a 19 gauge needle.
- Absolute alcohol.

With both agents, you should decide on the volume you need before you start the block. We recommend that you use 0.5 ml of solution per root level you wish to obtund; it is probably best not to use more than 2 ml at one session, although you may have to. Always be wary of using a large volume of alcohol, since it tends to spread widely.

Inject smoothly and slowly, and if you are using a relatively large volume, aspirate several times to make sure that your needle is still in the subarachnoid space, but do not barbotage.

You should, whenever possible, use a C-arm image intensifier to check the spinal levels when injecting neurolytic agents, but this is not always practical, since quite often the patient is very ill and the procedure has to be carried out on the bed.

Sacral Root Blockade

This route is indicated for perineal and/or pelvic pain; the agent of choice is phenol in glycerin.

If the patient is incontinent, sit him up as you would for a 'saddle block'.

If the patient is continent, try hard to avoid damage to the first and second sacral nerve roots, as they are the main nerves to the bladder: in this case use the Mehta position, where the patient sits up and leans back at 45 degrees and is tilted towards the side of the pain at 45 degrees. If the pain is bilateral or midline, it is often only necessary to ablate one side. In any case, do one side on one day and the other side 2 days later, thus minimising the possible damage to the first and second sacral nerve roots.

Because of the slope of the sacrum, it is better to carry out the block at the L3/4 or L4/5 space rather than at L5/S1, as you are less likely to drip phenol onto the upper sacral nerve roots (Figure 5).

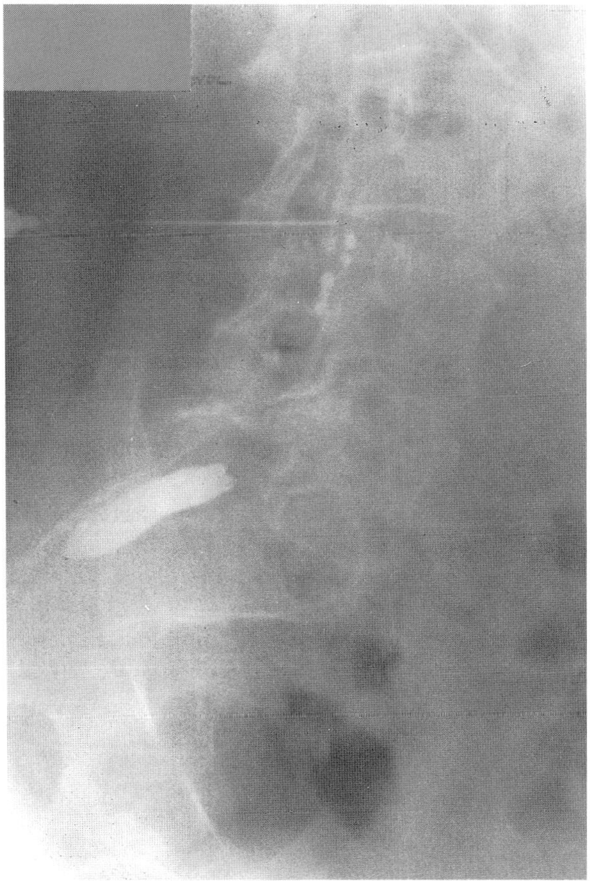

Figure 5. Intrathecal Phenol injection in to the lower sacral roots.

The patient should be left in position 30 min after the block.

2 Paravertebral Phenol Blockade

This technique is very useful when dealing with unilateral chest pain which covers several segments, thus making intercostal nerve blocks impractical; it can also be used as an alternative to intrathecal phenol, as subarachnoid blocks are technically more difficult at the thoracic level.

Technique

You can have the patient sitting up or lying down; it may be advisable to use a C-arm image intensifier to check your position, since you are injecting a neurolytic agent.

Place your middle finger over the spinous process of the vertebra situated in the centre of the painful area and then place your index finger next to it; introduce your needle attached to a syringe next to the index finger at the level of the cephalad leading edge of the spinous process (Figure 1). Your needle, vertically introduced, will hit either transverse process or lamina; when his has occurred, withdraw the needle slightly and reintroduce it at a more cephalad angle so that it enters the paravertebral space (Figure 2). It is advisable to maintain constant pressure on the barrel of the syringe as the needle is introduced; in the cases where the needle fails to encounter bone, you will feel a definite loss of resistance as the needle penetrates the costotransvere ligament to enter the paravertebral space. In this situation there is naturally no need to reangle the needle, as its tip is in the paravertebral space.

We recommend that you use a large-gauge needle, preferably a 19 gauge 46 mm; this allows for easy injec-

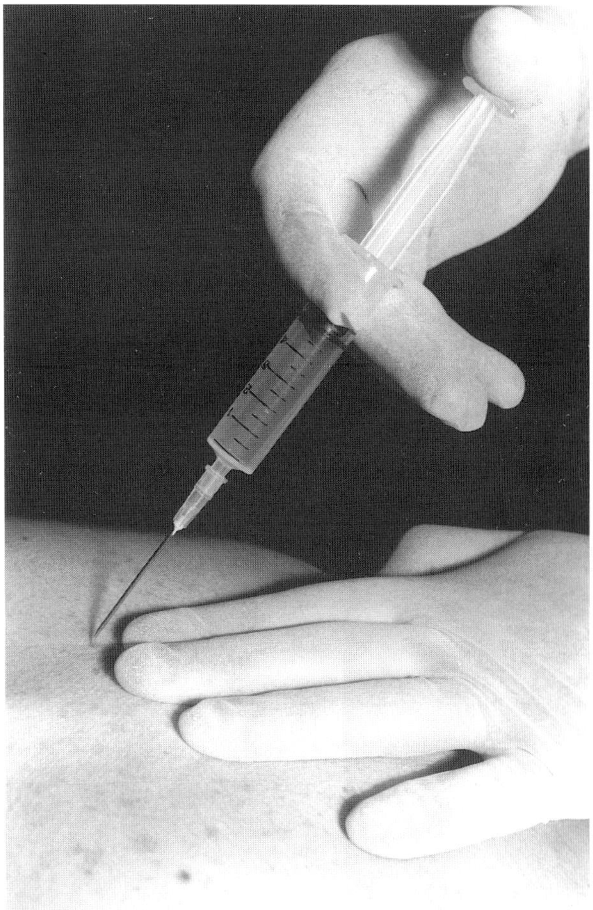

Figure 1. Introducing the needle for paravertebral blockade.

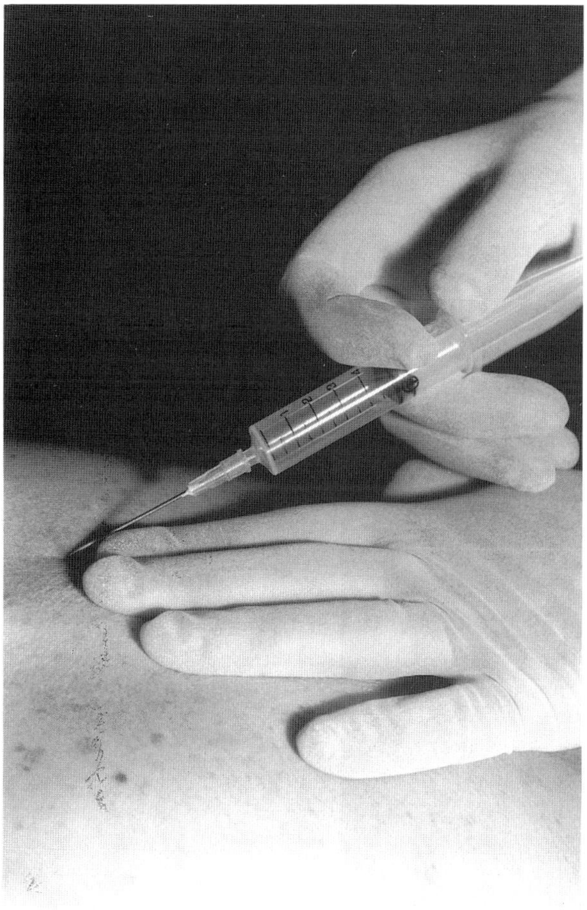

Figure 2. Reintroduction of the needle at a more cephlad angle to enter the paravertebral space.

tion of the phenol and also allows you more readily to appreciate the loss of resistance to injection.

Agent

We recommend that you use 1 ml of 7.5% phenol in 50% water and 50% glycerin per segment you wish to cover and that you enter the space in the middle of the painful area.

After-care

There is no need to leave the patient in a fixed position for long after injection, since the phenol in glycerin and water spreads very rapidly in the paravertebral space.

3 Other Sites for Neurolytic Agents

Epidural Space

Epidural phenol injection is not a very commonly used technique, as it is very difficult to be certain of the degree and direction of spread of phenol in the epidural space.

Peripheral Nerves

The only peripheral nerves the authors regularly use neurolytic agents on are the intercostal nerves; we would recommend the use of 1–2 ml of 7.5% phenol in 50% glycerin and 50% water per nerve. Neurolysis of peripheral nerves inevitably leads to deafferentation pain should the patient live long enough. The technique is therefore only recommended for patients with pain from a malignant cause who have a limited life expectancy.

4

Autonomic Blocks

1 Cervical Sympathetic Block (Stellate Ganglion Block)

Indication

Pain in the head, neck and upper limb mediated by the sympathetic system.

Technique

Sedation
Sedation is not usually necessary.

Position
The patient lies supine; his head is moderately extended so as to expose the principal landmarks in the neck (see below) (Figure 1).

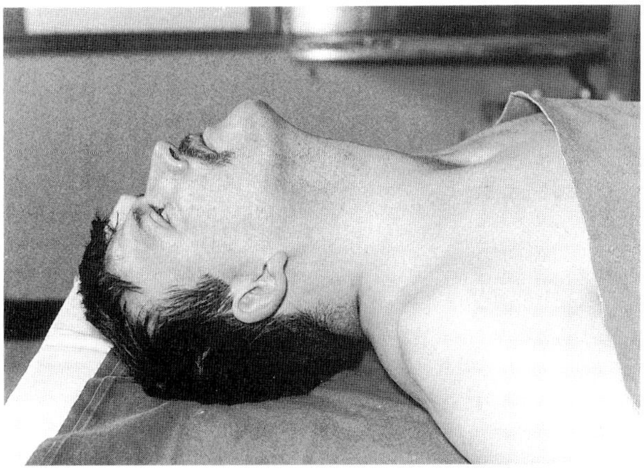

Figure 1. Position of patient for cervical sympathetic block.

Landmarks
You need to locate the most easily palpated cervical transverse process by inserting your index and middle fingers into the space between the sternomastoid and trachea. A good working rule to locate this transverse process is to flex your thumb at the interphalangeal joint and to position the joint in the patient's sternal notch with your thumbnail overlying the patient's trachea; using your thumbnail as a pivot, rotate your thumb laterally. Your interphalangeal joint should now overlie the most prominent cervical transverse process (Figure 2). If you now use your index finger and forefinger in the space described above, you should easily locate the spot, which lies lateral to the cricoid cartilage.

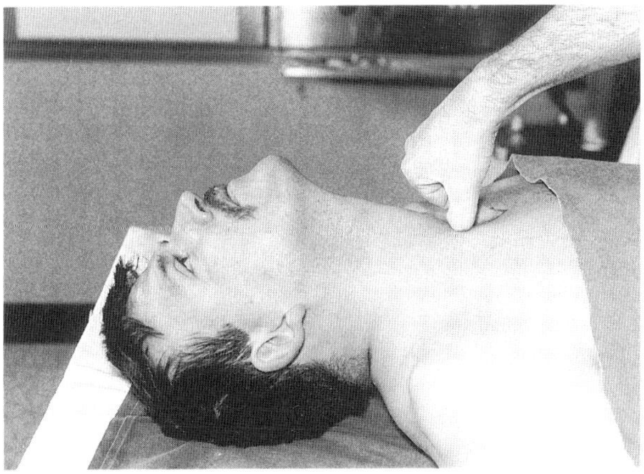

Figure 2. Locating the transverse process of the 6th. cervical vertebra.

Clean and drape the area.

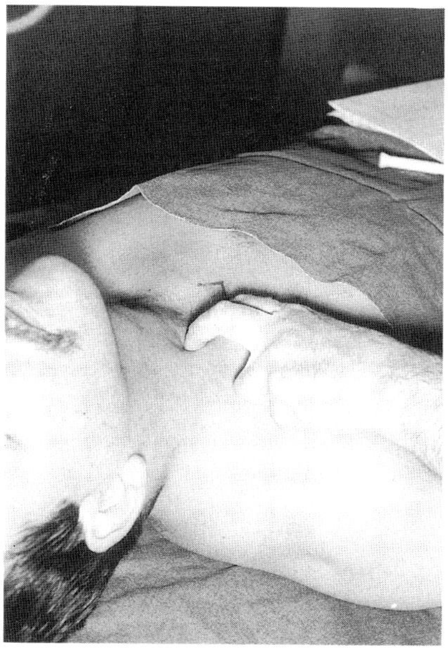

Figure 3. Pushing the carotid sheath laterally.

Procedure

Having located the spot, use your fingers to push the carotid sheath laterally 'out of harm's way' (Figure 3). Then insert vertically downwards a 21 gauge 40 mm or a 23 gauge 25 mm needle (according to the patient's build) attached to your syringe, so that it strikes bone; withdraw the needle slightly, fix it in that position, aspirate and inject (Figure 4). The authors inject 10 ml 0.25% plain Bupivacaine.

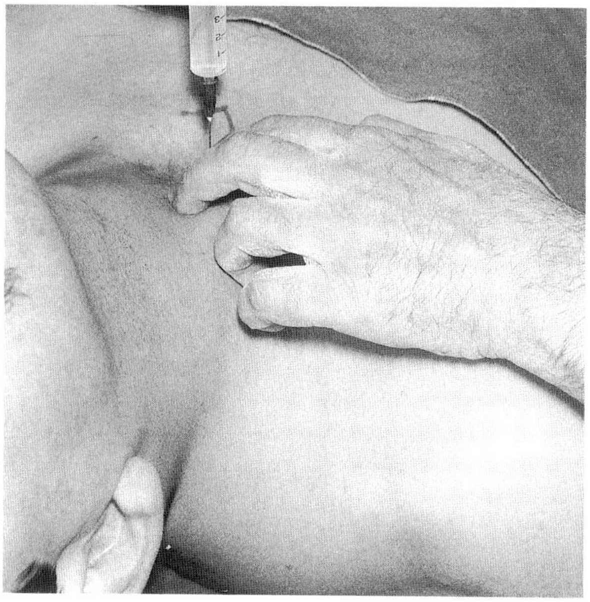

Figure 4. Injecting the local anaesthetic solution.

It facilitates matters if you ask the patient to open his or her mouth widely during the procedure and to mouth breathe: this prevents swallowing. Apart from making matters technically easier if the patient does not swallow, there is is also an important safety consideration to bear in mind: your needle may enter the oesophagus, and if the patient should swallow in this situation, a linear tear of the oesophagus could occur and this in turn could result in mediastinitis.

Once you have injected, sit the patient up for about 5 min.

Caution
There must be no resistance to your needle until you strike bone; if you meet resistance, stop, withdraw and start again.

Always aspirate, as an intravascular injection at this level will almost certainly produce severe convulsions. On the left side it is always possible to enter the thoracic duct; in this situation, aspiration will result in lymph entering the syringe.

There is always a theoretical risk of injecting through a dural sleeve into the cerebrospinal fluid; aspiration does not always result in cerebrospinal fluid appearing in the syringe, owing to the small size of the sleeve. The potentially lethal consequences of such an injection can be minimised by injecting the local anaesthetic in small aliquots, e.g. 2.5 ml.

Pneumothorax can occur even in the most carefully executed blocks!

2 Thoracic Chemical Sympathectomy

Indication

Pain in the thoracic region mediated by the sympathetic system and which is not responding to more conservative measures; one particular indication in the past was crippling anginal pain, although this has now probably been superseded by spinal cord stimulation (q.v.). An initial block with Bupivacaine would have given excellent pain relief.

Technique

Sedation

Sedation is not usually necessary.

Position

The patient lies prone on a radiolucent table. With upper thoracic work it is advisable to arrange the patient's arms so that they hang over the top edge of the table; this manoeuvre lifts the scapulae out of the way (Figure 1). We have sometimes opted for carrying out this block with the patient sitting up, as this has given us better anteroposterior and lateral views when using the C-arm image intensifier.

Landmarks

For anginal pain you need to block T1–T5. With the C-arm image intensifier screening in the anteroposterior plane, locate the spinous process of the second dorsal vertebra and mark a point 3.5 cm ($1\frac{1}{2}$ in, three fingers' breadth) lateral to it.

Clean and drape the area.

Procedure

Infiltrate the entry point with 0.5% Lignocaine with Adrenaline, angling your needle towards the body of the vertebra.

To carry out the block you should use a 9 cm 20 gauge needle with Pitkin point – i.e. blunt-tipped to minimise the risk of pneumothorax (Figure 2). The needle is inserted at an angle so as to hit the side of the vertebral body (Figure 3); confirm this by screening with the C-arm image intensifier. Always ensure that the notch on

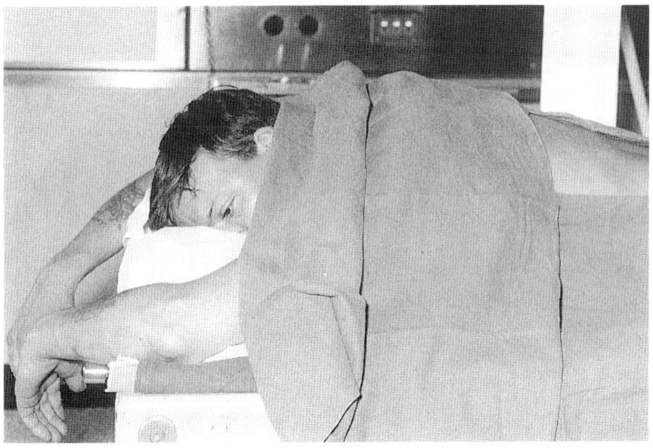

Figure 1. Position of patient for thoracic chemical sympathectomy.

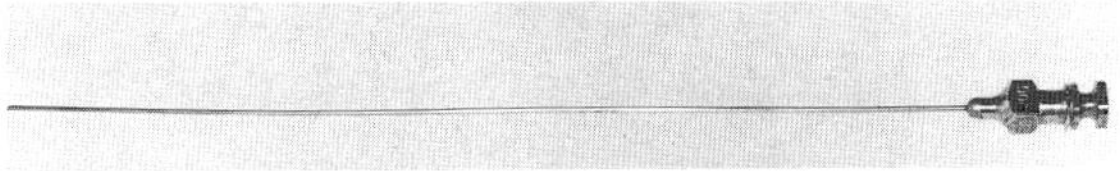

Figure 2. Needle used for thoracic chemical sympathectomy (Pitkin Point).

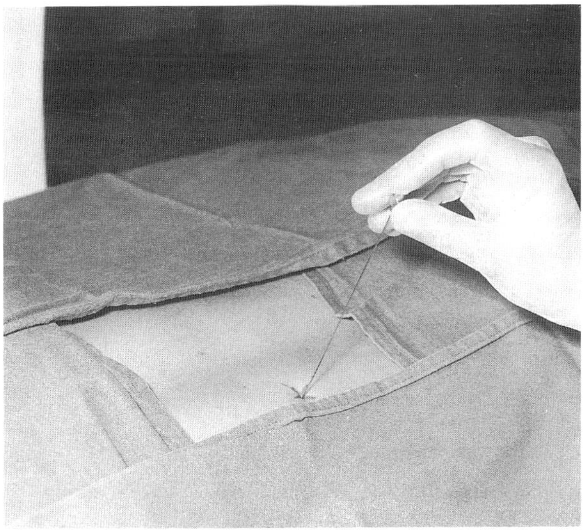

Figure 3. Inserting the needle.

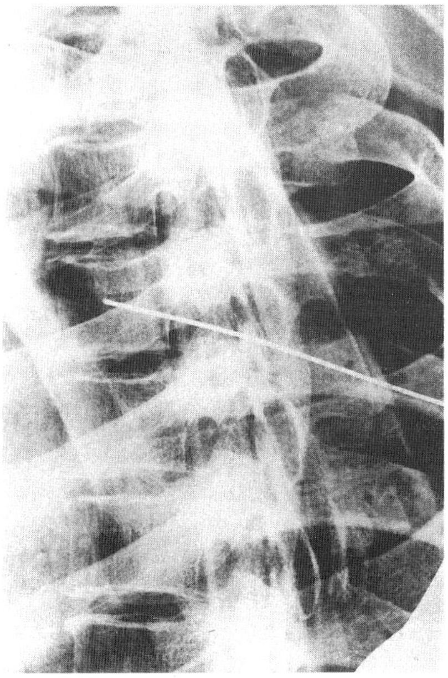

Figure 4. Lateral view of thoracic spine-needle in contact with body of thoracic vertebra.

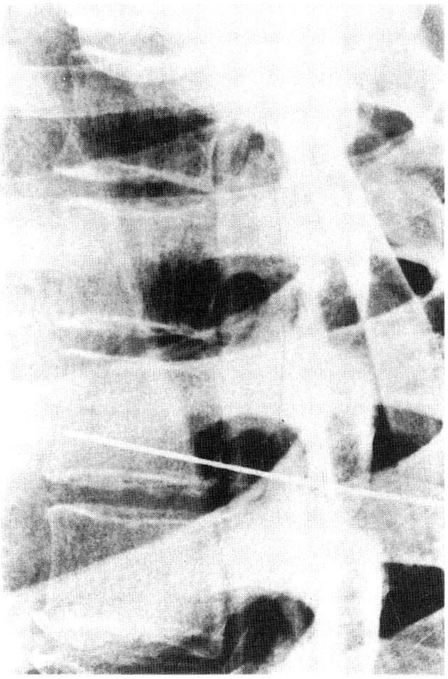

Figure 5. Lateral view of thoracic spine-needle in its final position.

the needle hub which indicates the side of the bevel tip is facing towards the vertebral body; this will prevent the bevel tip from digging into the periosteum, producing pain.

Rotate the C-arm image intensifier to screen in the lateral plane and manipulate it so as to get the best possible view of the side of the vertebral body (Figure 4). Under image intensifier X-ray control adjust the angle of the needle so that it now runs parallel to the body of the vertebra, and advance the needle very gently to a point 2–3 mm *behind* the edge of the vertebral body (Figure 5).

You should now confirm the position of your needle tip by injecting 2 ml of contrast medium – this will sometimes demonstrate the back of the pleura!

If you are satisfied with the position of your needle, inject 2 ml of 6% phenol in glycerin and water.

As soon as the injection has been done, put the patient in the prone position and keep him or her there for about 30 min so as to avoid posterior spread of the neurolytic solution onto the corresponding spinal nerve root.

> **Warning**
> This procedure carries a very real risk of pneumothorax, so proceed with care.

3 Lumbar Chemical Sympathectomy

Indication

Pain in the lower limbs mediated by the sympathetic system. The authors would advise caution in using this technique for non-ischaemic pain of benign aetiology. In both sexes there is the risk of producing genitofemoral neuralgia. In addition, in women, ankle oedema may result. A particular indication is severe ischaemic pain in elderly patients; it is of special benefit in rest pain and when skin nutrition has been endangered by indifferent blood flow. An initial block with Bupivacaine, when considered necessary, would have given excellent pain relief.

Technique

Sedation

Sedation is not usually necessary.

Position

- Position 1: The patient lies prone on a radiolucent table with a pillow under the abdomen in order to straighten the spine.
- Position 2: The patient lies in the lateral position on a radiolucent table with the affected side uppermost and the back to the operator.

In both cases you will need anteroposterior and lateral views of the vertebral column to ascertain the position of your needle. The angle of the C-arm image intensifier *vis-à-vis* the patient will, of course, depend upon whether the patient is lying in the prone or in the lateral position. The prone position is obviously preferred for bilateral blocks.

Landmarks

With the patient in either position, the C-arm image intensifier is positioned for screening in a vertical axis; the vertebral levels in question are located. These will depend upon whether you wish to block at a single level or at multiple levels. Place a mark about four fingers' breadth lateral to the vertebral spinous process(es).

Clean and drape the area.

Procedure

- Infiltrate the entry point with 0.5% Lignocaine with Adrenaline; then direct a 2 in needle towards the vertebral body and infiltrate.

- With a small scalpel blade make a small incision in the skin; take a 20 gauge 15 cm needle and insert it through the incision at an angle of about 15 degrees off the perpendicular and advance it medially and ventrally until it strikes the side of the vertebral body (Figure 1). Confirm this by screening with the C-arm image intensifier. Always ensure that the notch on the needle hub which indicates the side of the bevel tip is facing towards the vertebral body; this will prevent the bevel tip from digging into the periosteum, producing pain.
- Reangle the needle so that it slides off the side of the vertebral body and advance it gently until the needle tip appears to be just medial to the lateral margin of the vertebral body on radiological screening in the anteroposterior plane (Figure 2).
- Position the C-arm image intensifier to screen in the lateral plane and gently advance the needle so that the tip comes to lie just posterior to the anterior border of the vertebral body (Figure 3). Very often you can elicit a definite 'give' as the tip of your needle passes through the psoas sheath to enter the paracolic gutter.
- Having ascertained that the patient is not allergic to iodine, inject 0.5 ml of a water-soluble non-ionic radio-opaque dye through the needle. If the needle is correctly positioned, the dye should enter the

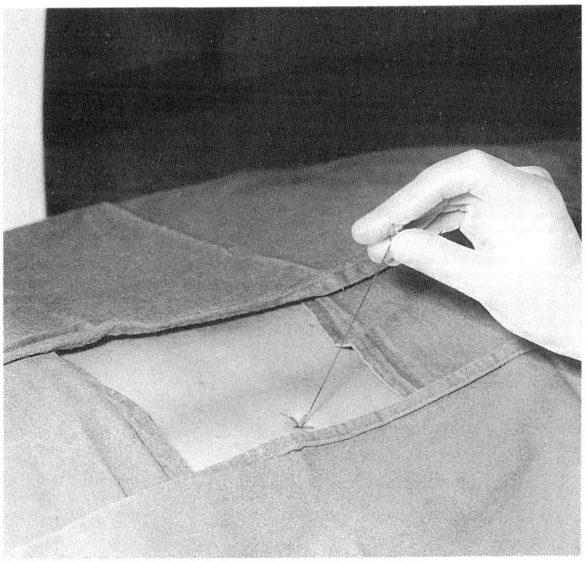

Figure 1. Inserting the needle.

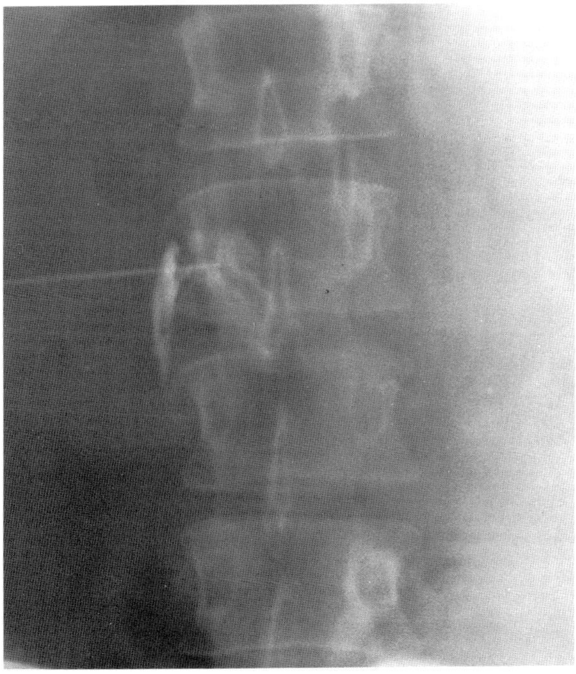

Figure 2. Postero/Anterior view of lumbar spine with needle tip in position: note the presence of contrast medium outside the paravertebral gutter. The needle had been repositioned!

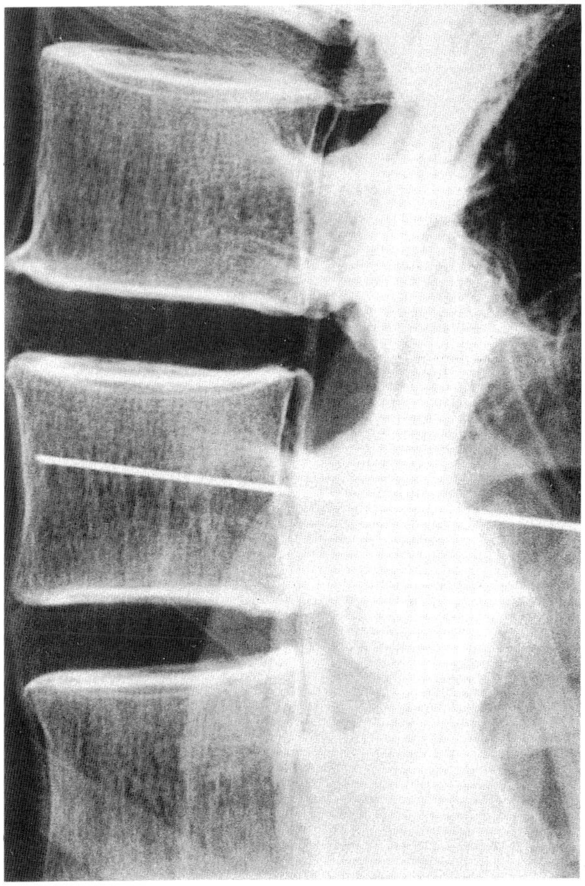

Figure 3. Lateral view of lumbar spine with needle tip in position.

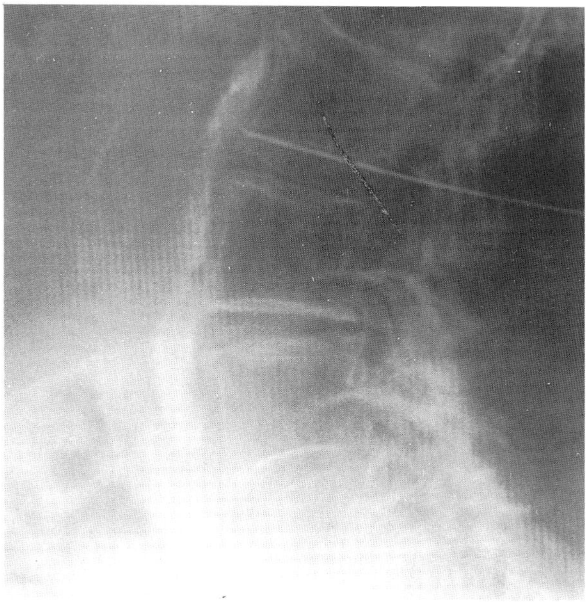

Figure 4. Lateral view of lumbar spine with needle tip in position; contrast medium in paravertebral gutter.

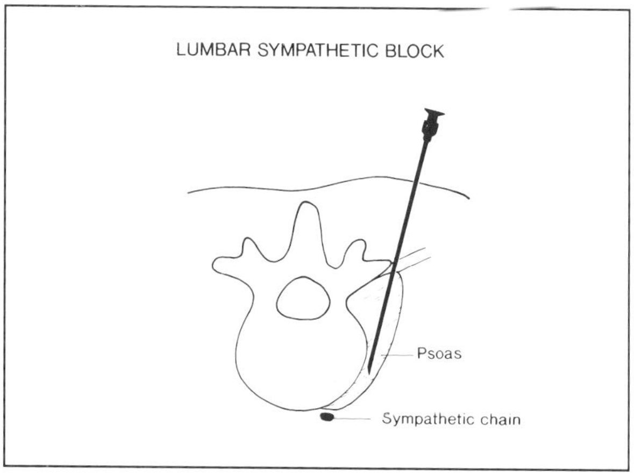

Figure 5. Schematic representation of needle tip still within psoas sheath.

paracolic gutter and spread both cephalad and caudad, as shown in Figure 4. Reposition the image intensifier to screen in the anteroposterior plane so as to ensure that there is no lateral spread of the contrast medium; if this is seen, it means that the tip of your needle is still within the psoas sheath and it needs to be advanced ventrally (Figure 5). Very muscular patients often have bulging psoas muscles, and in these cases you may find that the tip of your needle may have to be inserted as far as the anterior edge of the vertebral body or just beyond the anterior edge

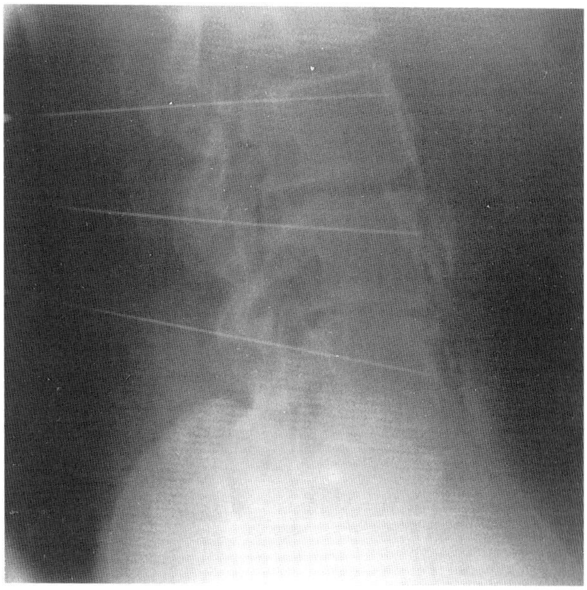

Figure 6. The 'three needle technique'.

before you can get a suitable spread of the radio-opaque dye.

- If you are satisfied with the radiological picture, you are now ready to inject your agent. In a 'trial block', assuming you are using a single-needle technique, the authors recommend 10 ml of 0.25% plain Bupivacaine. In a therapeutic lytic block, using a single-needle technique, inject up to 10 ml of 6% phenol in water. With a single-needle technique, there is less control over the spread of the relatively large volume injected at one site; there must theoretically exist, therefore, an increased risk of lateral spread of agent, with production of genitofemoral neuralgia. An alternative is to introduce three needles at the L2, L3 and L4 levels and to inject 2–3 ml of agent through each needle (Figure 6). It is always safer to inject smaller volumes of a neurolytic agent. A relatively recent trend has been to insert a single needle at the L3 level and to inject 2–3 ml of agent through it.

Remember to aspirate before any injections, as there are large blood vessels about!

After-care

It is prudent to keep the patient lying in the lateral position with the affected side uppermost for about 30 min following a unilateral lytic block, so as to keep the phenol away from the genitofemoral nerve; this minimises the risk of genitofemoral neuralgia.

4 Coeliac Plexus Block

Indication

Pain originating from the abdominal viscera and which is not responding to more conservative measures.

Technique

Before the patient is brought to theatre, ensure that he or she is wearing elasticated stockings. Set up an intravenous infusion with clear fluid.

Sedation

Sedation is invariably necessary.

Position

The patient lies prone on a radiolucent table, arms forward above the head; place a pillow under the abdomen so as to straighten the lumbar spine (Figure 1).

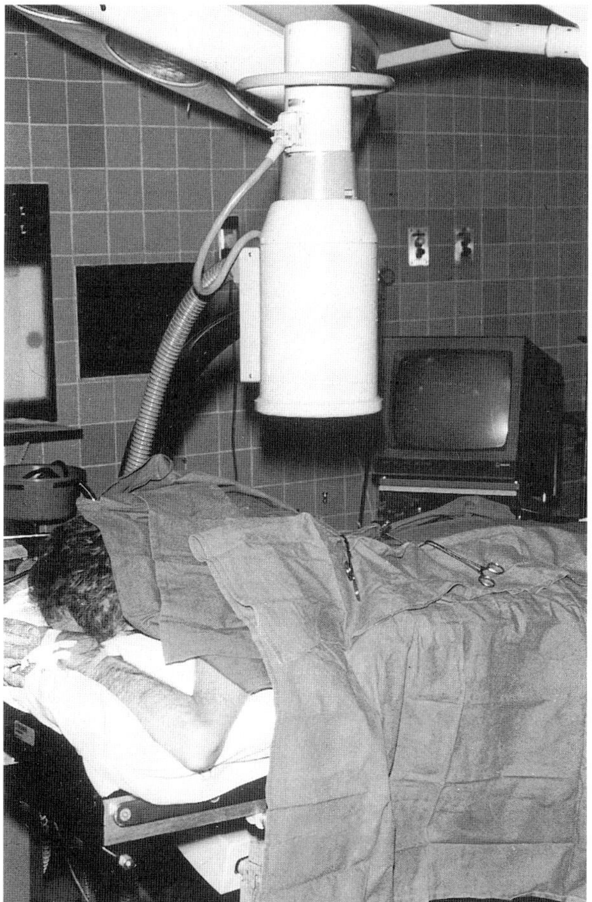

Figure 1. Position of patient for coeliac plexus block.

Landmarks

With the C-arm image intensifier screening in the anteroposterior plane:

- Locate the spinous process of L1 and mark its cephalad leading edge.
- Draw a line about four fingers' breadth lateral to the

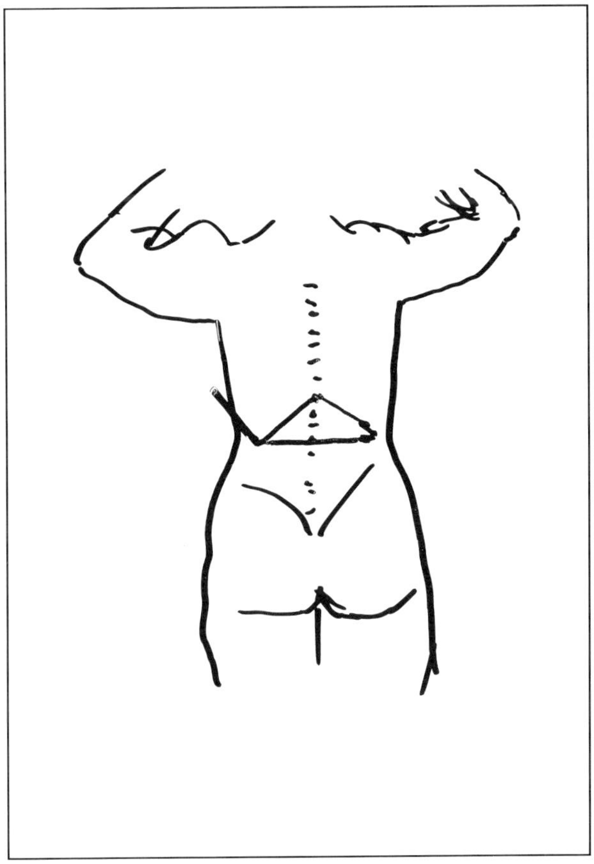

Figures 2a and 2b. Landmarks for coeliac plexus block.

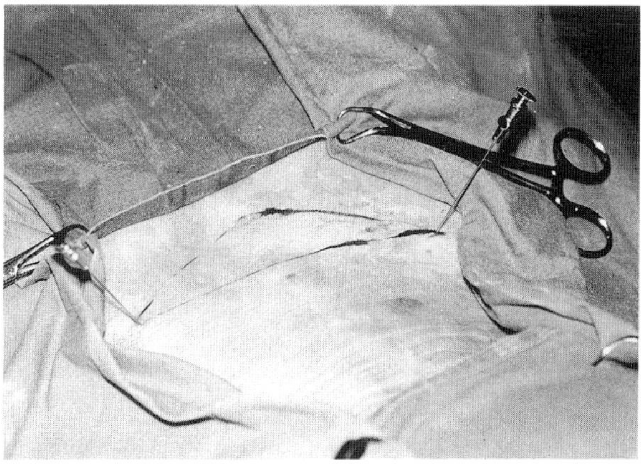

mark on each side; make a second mark (on each side) where the line crosses the 12th rib.

- Join the three marks together so as to construct a triangle (Figure 2a, b). The lateral angles of this triangle constitute your entry points.

Clean and drape the area.

Procedure

- Infiltrate the entry point with 0.5% lignocaine with adrenaline, then direct a 5 cm (2 in) needle towards the vertebral body of L1 and infiltrate.
- With a small scalpel blade make a small incision in the skin. Take a 19 gauge 15 cm needle and insert it through the incision; owing to the natural curvature of the body, it will be found possible to insert your needle at right angles to both lines forming the lateral angle (Figure 2b). This is the correct angle at which to advance your needle. The path it follows will be ventrally, medially and cranially until it strikes the side of the first vertebral body; confirm this by screening with the image intensifier. Always ensure that the notch on the needle hub which indicates the side of the bevel tip is facing towards the vertebral body; this will prevent the bevel tip from digging into the periosteum, producing pain.
- Reangle the needle so that it slides off the side of the vertebral body and advance it gently until the needle tip appears to be well medial to the margin of the vertebral body on radiological screening in the anteroposterior axis (Figure 3).
- Position the C-arm image intensifier to screen in the lateral axis and gently advance the needle so that the tip comes to lie just ventral to the anterior border of the vertebral body (Figure 4).
- Having ascertained that the patient is not allergic to iodine, inject 0.5 ml of a water-soluble non-ionic radio-opaque dye through the needle. If the needle is correctly positioned, the dye should enter the paracolic gutter and spread both cephalad and caudad.

Reposition the image intensifier to screen in the anteroposterior plane so as to ensure that there is no lateral spread of the contrast medium; if this is seen, it means that the tip of your needle is still within the psoas sheath and needs to be advanced ventrally (Figure 5).

After positioning the needle on one side, repeat the procedure for the opposite side.

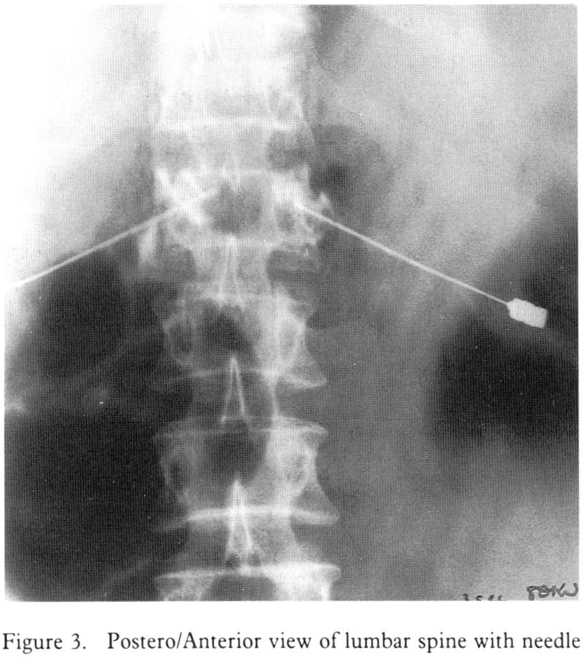

Figure 3. Postero/Anterior view of lumbar spine with needle tips in position.

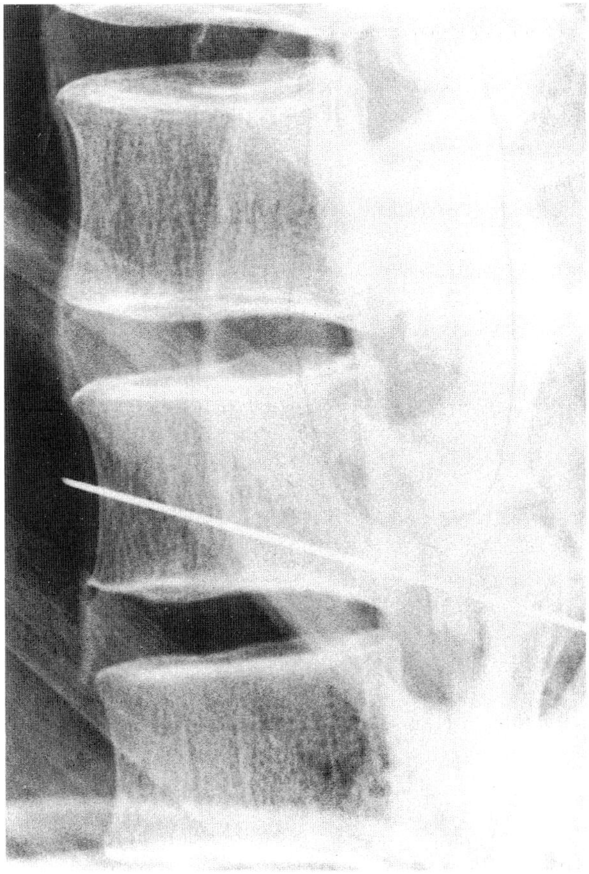

Figure 4. Lateral view of lumbar spine with needle tip in position.

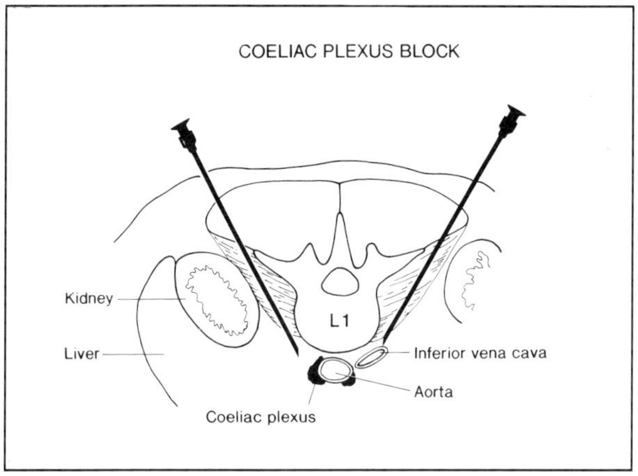

Figure 5. Schematic representation of final needle positions in coeliac plexus block.

- You are now ready to inject your agent. In a 'trial block', we recommend 20 ml of 0.25% plain Bupivacaine down each needle. In a therapeutic lytic block, we suggest that you first inject 2.5 ml of 2% Lignocaine on each side and wait for a few minutes; then inject 10 ml of 0.5% Bupivacaine plus 10 ml of absolute alcohol on each side.

  Remember to aspirate before any injections, as there are large blood vessels about!

After-care

Keep the patient in bed for the rest of the day and monitor the blood pressure; hypotension may need to be counteracted by elevating the foot of the bed and by administering intravenous fluids. If the blood pressure is satisfactory the next morning with the patient supine, ask him or her to stand erect, and measure the blood pressure again so as to ascertain that there is no postural hypotension. The elasticated stockings should be kept on until you are happy that the blood pressure has stabilised. The intravenous fluid infusion should also be kept going until stabilisation of the blood pressure has been achieved.

5 Intravenous Sympathetic Block

Indication

Pain in the limbs mediated by the sympathetic system.

Technique

Sedation

Sedation is not usually necessary.

Position

The patient lies supine.

Procedure

- An intravenous cannula is inserted into the dorsum of the hand or foot, depending in which limb is to be blocked. Venous access is also secured by means of a cannula in the (other) hand as a precaution against accidental cuff deflation. Exsanguination of the limb must then be carried out, either by using an Esmarch bandage or by elevating the limb for about 5 min. In the case of an arm block, pressure may be exerted over the brachial artery while the patient is asked to exercise the elevated hand. As this is being done, an assistant attaches an orthopaedic tourniquet around the thigh or upper arm; at the end of the exsanguination, the tourniquet is inflated to about 50 mm/Hg above the patient's normal systolic blood pressure.
- You should now inject your mixture through the intravenous cannula in the limb to be blocked (Figure 1). The most important component of the mixture is

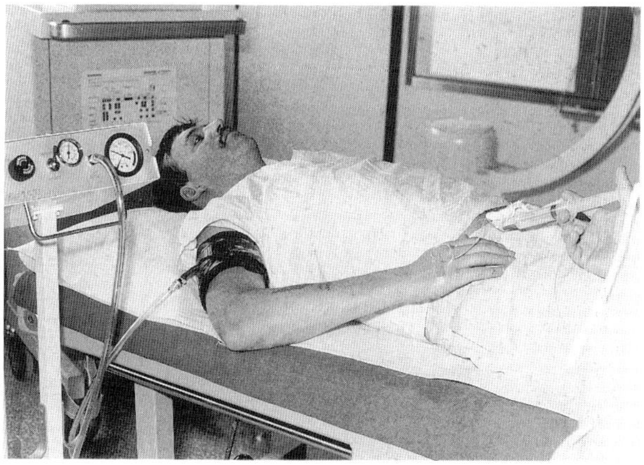

Figure 1. Intravenous sympathetic block.

Guanethidine; use up to 20 mg of the drug, depending on the age and cardiovascular state of the patient. The vehicle for injecting the drug is usually:

Prilocaine, a very safe agent to use intravenously. Use the 0.5% solution (without preservative). As with saline, use up to 50 ml of solution, depending upon which limb you are blocking and upon the patient's build.

- There is always a degree of leakage of solution from beneath the cuff, and the presence of a local anaesthetic in the mixture makes the procedure less uncomfortable for the patient.

  Some workers still advocate the addition of heparin, 1–2 ml of 1/1000, in order to minimise the risk of deep vein thrombosis.
- Keep the tourniquet in place for 20 min. At the end of that time deflate the tourniquet and reinflate it a few times so that the Guanethidine escapes into the general circulation gradually.

Note
You may need to carry out several of these blocks.

Upon release of the tourniquet some patients may complain of feeling faint or 'funny', and some may complain of a throbbing headache or of an unpleasant taste; these effects are transitory. It is quite common for patients to develop a headache or bloodshot eyes the day after the block. If these symptoms are particularly unpleasant, you may have to reduce the dose of guanethidine. It is not at all uncommon for young men to complain of impotence after several blocks over a short period of time. In this situation reassure the patient and stop the blocks! This complication will soon disappear.

5

Pituitary Ablation

Indication

Pain from multiple malignant secondary deposits in bone and which is not adequately controlled by analgesics. Pain from bone secondaries caused by carcinoma of the breast or carcinoma of the prostate is the most responsive to this technique.

Pre-operative Assessment

- The patient must be fit for general anaesthesia.
- A pre-operative lateral X-ray picture of the skull should be made; study carefully the position of the pituitary fossa and note the posterior clinoid processes (Figure 1).
- A pre-operative examination of the patient's visual fields should be carried out by an ophthalmologist, the record being filed in the clinical notes.
- To carry out this procedure, you will need a 15 gauge, 12.5 cm transphenoidal cannula with fitting trochar and a Kelsey Fry Mallet (Figure 2a–c).

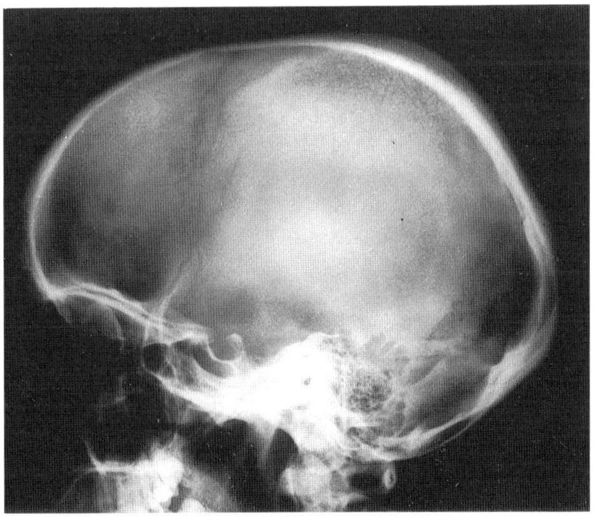

Figure 1. X-ray skull-lateral view to show pituitary fossa.

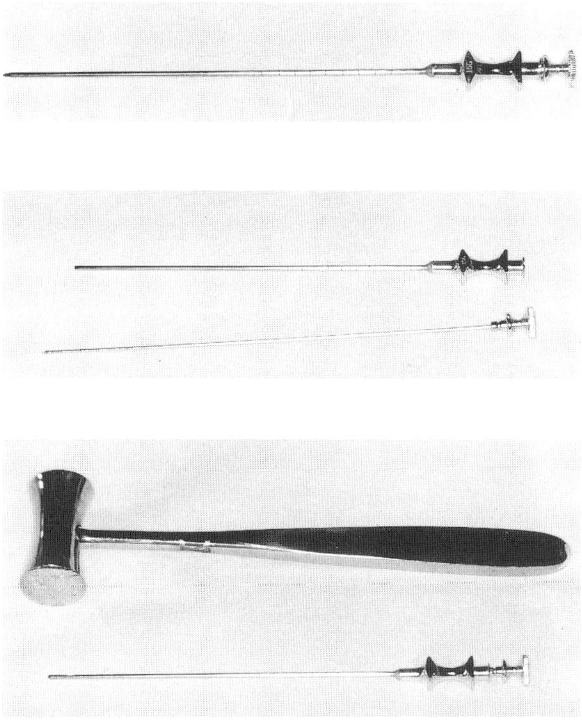

Figures 2a, 2b and 2c. Transphenoidal cannula, fitting trochar and Kelsey-Fry mallet.

1 Alcohol Ablation

Technique

Sedation

The procedure requires full general anaesthesia with oral endotracheal intubation; we recommend that you insert cocaine paste/drops into the patient's nostrils and that you also administer an intravenous antibiotic after induction; our practice is to use Ampicillin and Flucloxacillin in the non-penicillin-sensitive patient.

Position

The patient lies supine with the head on a radiolucent rest which affords full radiological access (Figure 3).

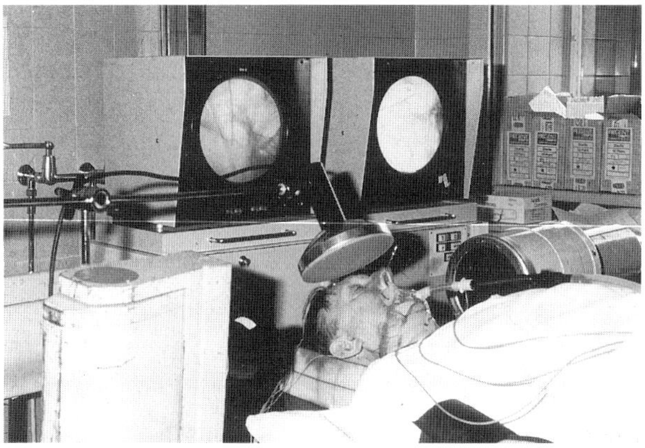

Figure 3. Position of patient for pituitary ablation.

Landmarks

Position the image intensifier so that it screens in the lateral plane (Figure 3); note the pituitary fossa and study carefully the posterior clinoid process. When screening in the lateral plane, you will be aiming your trochar and cannula to a point just below the apex of the process (Figure 4). In other words, the trochar tip should come to rest on bone and be overhung by the process.

Rotate the C-arm image intensifier so as to get a good anteroposterior view of the skull; when screening in this plane, your cannula must come to rest in the midline (Figure 5). Remember that on each side of the pituitary fossa lie the cavernous sinuses containing the internal carotid arteries.

Procedure

- Clean the face and the inside of each nostril.

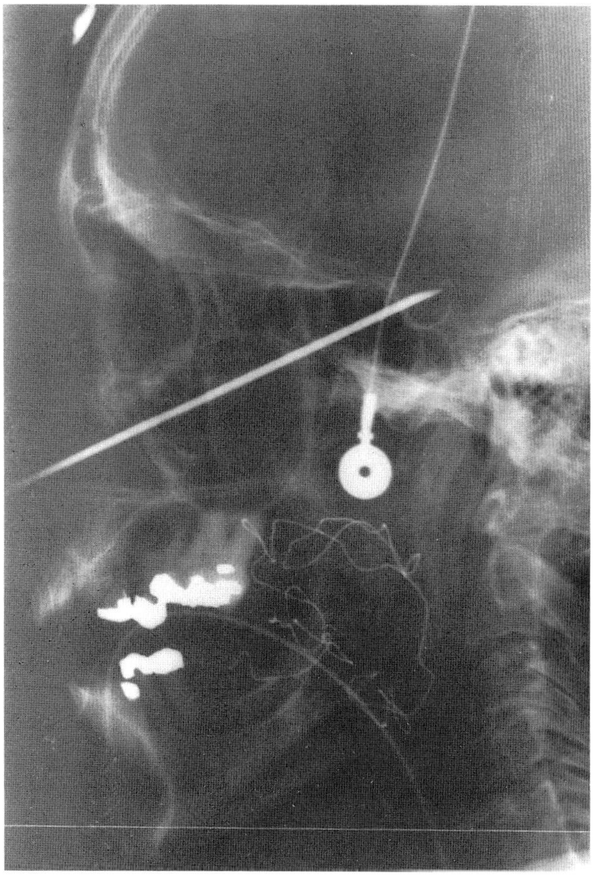

Figure 4. Lateral view of skull with trochar and cannula correctly positioned i.e. below posterior clinoid process.

- Stand behind the patient, holding the trochar and cannula; if you are right-handed, hold the trochar and cannula in your left hand and insert it in the patient's right nostril, aiming for the bridge of the nose in the anteroposterior plane and for an imaginary point in front of the tragus of the right ear and behind the right eye in the lateral plane. Push and twist the trochar in so that it is gripped by the soft tissues, and check your position radiologically in both planes as you direct it towards the posterior clinoid process.
- When you are satisfied with this preliminary position of the trochar and cannula, tap lightly with the mallet so that the trochar and cannula 'support themselves'. Continually checking the position of your trochar and cannula in both X-ray planes and reorientating as necessary, tap them further and further into the bone (Figure 6). A dramatic loss of resistance will indicate that you have entered the sphenoid sinus. Keep going and you will soon encounter hard bone again; this is now the front of the pituitary fossa. Check your position again and drive the trochar and cannula into the

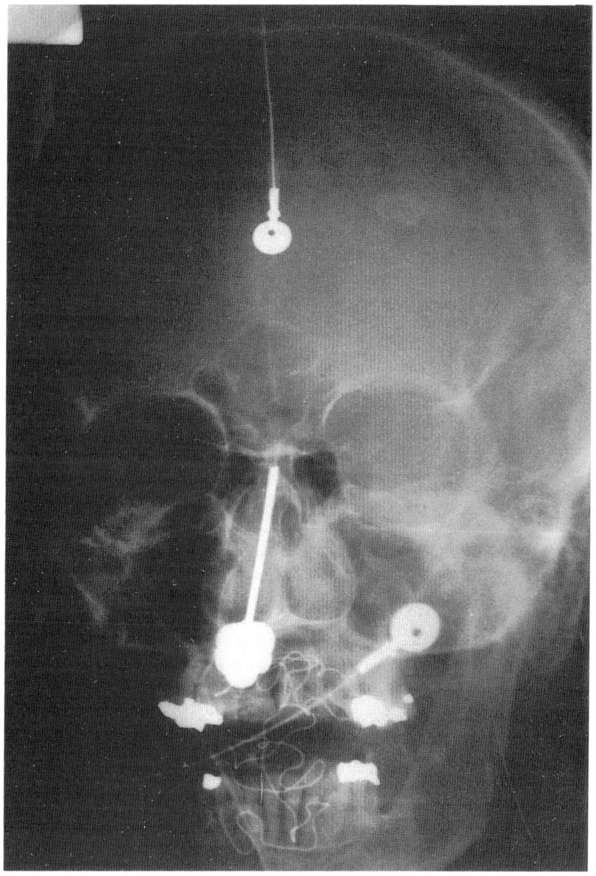

Figure 5. Antero/Posterior view of skull with trochar and cannula in correct midline position.

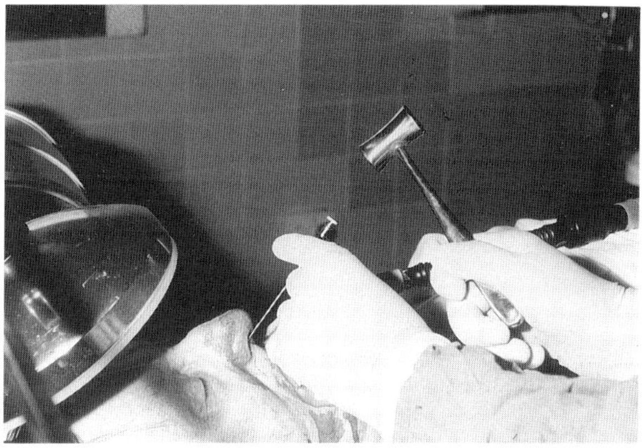

Figure 6. Tapping trochar and cannula into position.

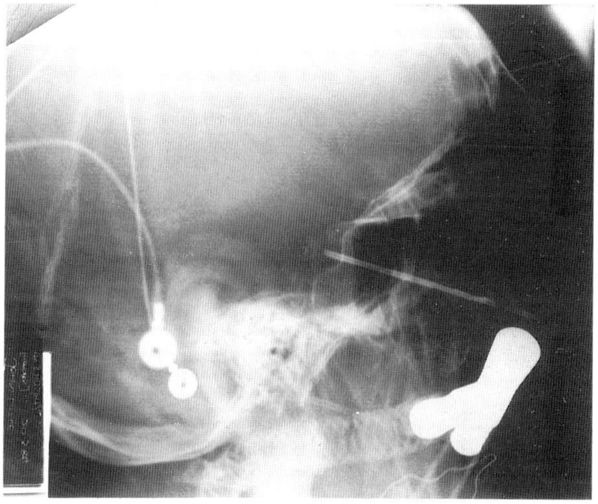

Figure 7. Trochar withdrawn leaving cannula in its correct final position.

fossa, aiming, as already stated, for a point just below the apex of the clinoid process in the lateral X-ray view and strictly in the midline in the anteroposterior view (Figures 4, 5).

- The trochar and cannula should be inserted to the very back of the pituitary fossa; when the trochar is withdrawn from the cannula, the tip of the cannula will then be just short of the bone at the back of the fossa (Figure 7). The aim is to deposit alcohol partly into the posterior part of the pituitary gland and partly into the anterior part of the gland.

We have found changes in pupillary size to be an erratic monitor of potential damage to the optic nerve; we strongly recommend the use of evoked potentials as a far safer way of monitoring changes in optic nerve function as a result of alcohol injection into the pituitary gland (see below).

The total dose of absolute alcohol used is 1 ml injected in 0.1 ml aliquots, observing pupillary size/evoked potentials as you do this (Figure 8a, b). The volume of the cannula employed by us is 0.3 ml, so the first three aliquots will consist entirely of air. It is essential that you find out the volume ('dead space') of your cannula.

Deposit 0.8 ml of syringe contents through the can-

Note

When the trochar is withdrawn, blood may ooze out of the needle. If the anteroposterior position of the needle is correct on radiological screening, this indicates that you have probably entered the pituitary plexus. In this case withdraw the needle into the centre of the pituitary fossa and redirect it slightly caudad, i.e. away from the apex of the posterior clinoid process and more towards the base of the fossa in the lateral view.

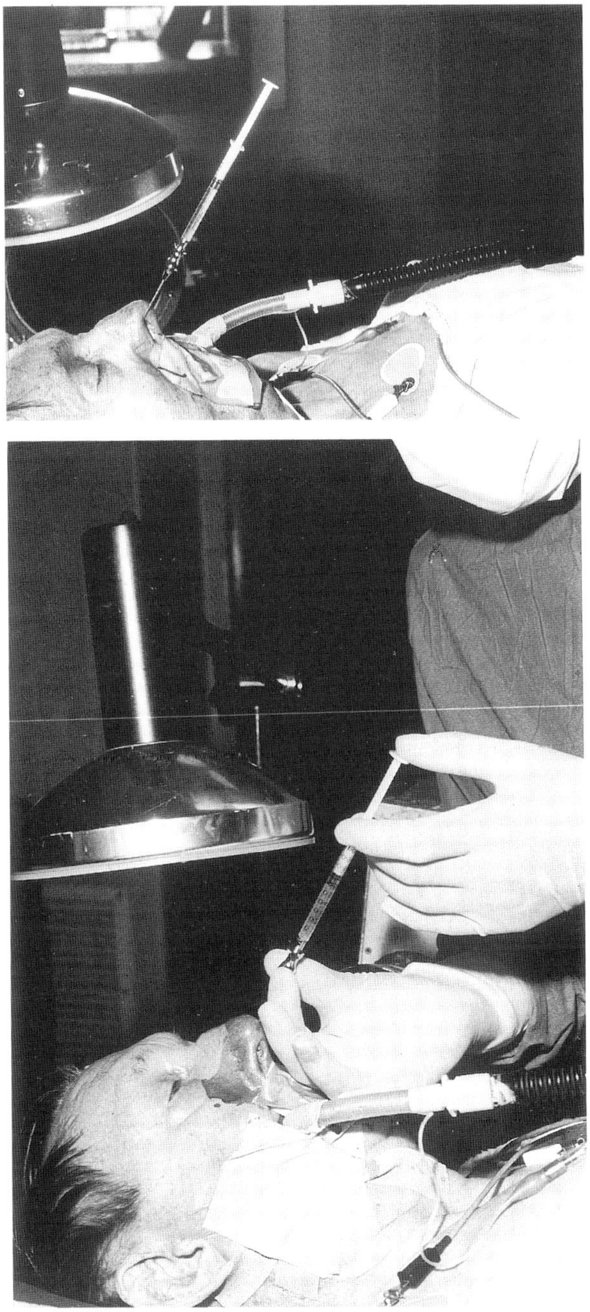

Figures 8a and 8b. The injection of alcohol.

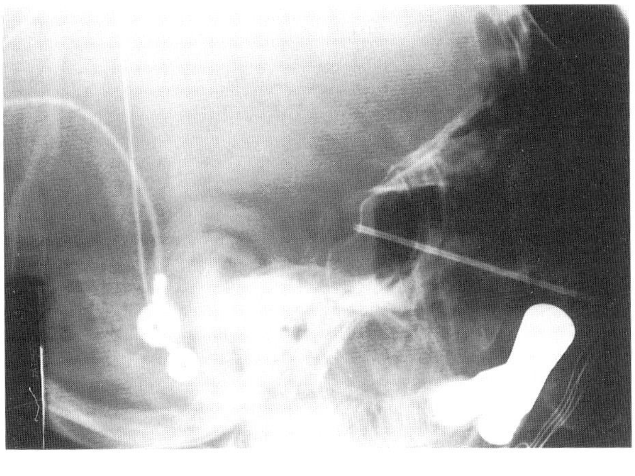

Figure 9. Cannula in anterior part of pituitary gland.

nula; this will mean that you have injected 0.3 ml of air plus 0.5 ml of alcohol into the posterior pituitary. Now withdraw the cannula to the anterior pituitary (Figure 9) and inject the rest of the syringe contents, i.e. 0.2 ml. This pushes 0.2 ml out of the cannula dead space into the pituitary and replaces that with the 0.2 ml of syringe contents which you have injected into the cannula. The trochar should now be gently inserted into the cannula in order to empty the contents of its dead space into the gland. The insertion of the trochar should be done in three stages, so that at each stage a 0.1 ml aliquot of cannula dead space contents is injected. The markings on the needle will guide you in this manoeuvre (Figure 2a).

Check evoked potentials/pupillary responses (i.e. presence of pupillary dilatation) after each aliquot is injected. If an abnormal response is produced, wait for 5 min and check again. If the changes persist, abandon the procedure. It is better to repeat the procedure a second time than to produce a visual handicap in an already incapacitated patient.

When you have carried out the injection withdraw the trochar and cannula.

Caution!
Aspiration of blood or cerebrospinal fluid means that you *must* resite your trochar and cannula.

2 Cryotherapy

In our experience monitoring pupillary reactions when injecting alcohol is insufficient to provide absolute protection against damage to the optic nerve. If you do not have the facility to monitor evoked potentials from the optic nerve, we suggest that you use the technique of pituitary cryoablation as described by Duthie (1983)(*).

The rationale of cryotherapy is dealt with in detail on page 35. Here use is made of a special cryoprobe constructed so as to pass through the 15 gauge 12.5 cm transphenoidal cannula. The technique described above is used to insert the cannula into the pituitary fossa and a series of four lesions ('freezes') are made. The cannula is then withdrawn and inserted through the other nostril into the other side of the fossa, and, following the same procedure, another four lesions are made. One week later the procedure is repeated.

After-care

Whatever technique you use, whether alcohol or cryotherapy, after-care is important. It is prudent to keep the patient in hospital for a few days post procedure; his or her visual fields will need to be tested at least once during the stay.

Fluid Intake

Through an intravenous cannula, administer Dextrose/saline 1 l over 24 h and monitor fluid balance; encourage drinking and do not overdo the intravenous fluids as you may precipitate a diuresis. The patient may exhibit a diuresis for a couple of days but this usually settles. If it does not settle, consider use of Pitressin.

Antibiotics

Keep on oral antibiotics for 5 days following the procedure.

NSAID

If the patient is not already on Non-steroidal Anti-inflammatory Drugs or steroids, give Indomethacin (slow-release preparation) orally 75 mg b.d. You can also use suppositories. These drugs reduce renal blood flow and sodium loss.

Analgesics

The patient should be maintained on their previous dose of analgesics for a few days; however, you should monitor respiration very carefully, in case the procedure rapidly produces analgesia and the patient's drug regimen becomes excessive to analgesic requirements.

Cerebrospinal Fluid Leak Through Nostril

Do not worry about this, as it usually settles spontaneously; if it proves to be severe and persistent, ask an ENT surgeon to consider inserting a muscle graft to plug the gap.

Visual Evoked Potentials

Evoked potentials are a standard neurological test for multiple sclerosis and are available in most hospitals through the electroencephalographic department.

One of the possible complications of injecting alcohol into the pituitary gland, due to the proximity of the optic nerve, is that of causing temporary or even permanent blindness. The traditional method of detecting incipient damage to the optic nerve is by observing the pupillary responses to the injection. However, this may be hindered by the effect of opiate analgesics. Another safer method is that of observing the visual evoked response, which is a very sensitive indicator of optic nerve function and which can be used to detect early changes in the conduction of the optic nerve.

Theory

If a visual stimulus is applied to the eye, a signal passes from the eye along the optic nerve to the brain. The amplitude of this signal is of the order of 10 μV and is normally impossible to detect because of the ongoing electroencephalographic (EEG) activity in the brain. If, however, a repetitive stimulus is applied, the response of the brain to the stimulus can be extracted from the ongoing activity by averaging the EEG signal following the stimulus. This technique of 'evoked potentials' has been widely used to study ophthalmic defects. The response to a patterned stimulus has been shown to be surprisingly constant between patients and to be extremely sensitive to subclinical lesions in the optic nerve. Unfortunately, during anaesthesia it is difficult to provide a patterned stimulus in a reproducible way and it is consequently necessary to use a flash of light as stimulus. However, the principal difficulty encountered with flash stimulation is the intersubject variability of the response. This is not a problem when monitoring the progress of potential visual impairment during an operation, since the patient can be used as his or her own control.

Technique

The response is recorded from two electrodes situated 1 cm and 5 cm anterior to the inion on the midline. Monopolar recording is used with the reference electrode placed on the right mastoid. The bipolar response between the two active electrodes can be obtained by subtracting the two monopolar responses, should this prove necessary (owing to higher than normal levels of background activity or noise). The stimulus is from a flash of light of intensity 1.5 J which is subtended approximately 35 degrees. Usually the response to 64 flashes is averaged, but occasionally it is necessary to increase this number. The amplitude, shape and latency of the response are monitored after each 0.1 ml of alcohol is injected into the pituitary fossa.

Unexplained reductions in response amplitude of more than 60%, a sustained increase in latency of more than 15 μs or a marked change in the shape of the response would be taken as indicating a hazardous situation. Increased caution would be considered necessary if changes greater than half the above values should occur.

(*) Duthie, A.M. (1983) Pituitary cryoablation. *Anaesthesia*, **38**, 495–497.

6

Implantable Epidural System

Indication

Usually for pain of malignant origin which cannot be controlled by reasonable oral dosage of opiates.

Technique

Equipment

The authors' experience has been gained using various systems. What follows is a general account of how to implant an epidural catheter and reservoir; there will be minor variations in technique from system to system. Figure 1 shows a typical implantable epidural system.

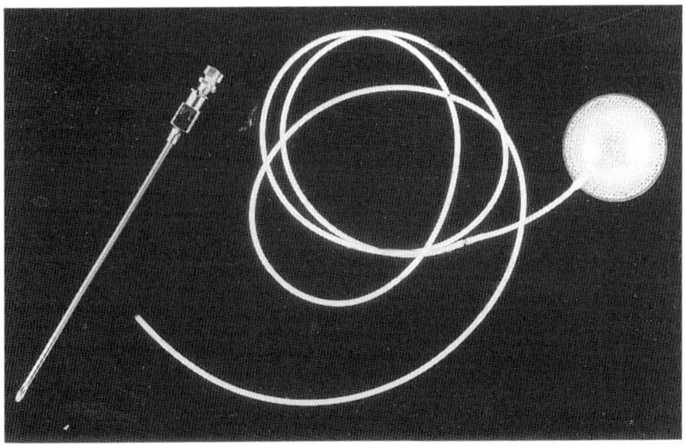

Figure 1. An implantable epidural system.

Sedation

Sedation is not usually necessary.

Antibiotic Cover

Antibiotic cover is recommended; we suggest Ampicillin and Flucloxacillin in a loading intravenous dose of 500 mg each with 250 mg of each orally every 6 h for 5 days.

Position

The patient lies in the lateral position, as he or she would for a normal epidural block.

Procedure

- Insert the Tuohy needle into the epidural space, using your preferred technique. When you are confident that you have entered the epidural space, take hold of the epidural catheter and feed it through the needle into the epidural space as you would any epidural catheter; some catheters are mounted on stiff stylets to aid introduction.

  You may want to use a C-arm image intensifier if you have decided to implant the catheter at a definite vertebral level; injection of a small dose of a suitable contrast medium, such as Omnipaque, through the catheter may aid its precise siting.

  If you do intend to implant at a definite level, remember that you will need to extract a length of catheter when you remove the Tuohy needle, so you will have to push it one or two segments above your target zone. If you experience difficulty in advancing the catheter past your desired level, you must take great care when removing the catheter; in these cases try to advance the catheter as you remove the Tuohy needle.
- When you are happy with the final position of the catheter, cut down onto the Tuohy needle until you

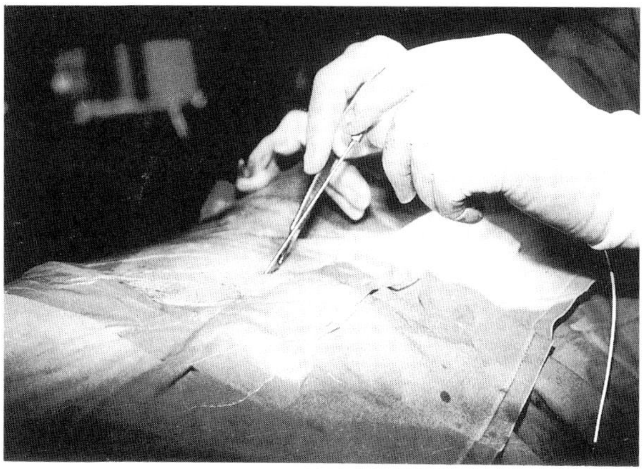

Figure 2. Cutting down onto Tuohy needle.

reach tough fascia (Figure 2). Remove the Tuohy needle, taking great care not to pull out the catheter; proceed in the same way as you would in removing any Tuohy needle containing an epidural catheter.

- The next stage is to plan out the site for the epidural reservoir. The best site is on the non-dominant side over a hard surface, e.g. the rib cage or the iliac crest; this allows the patient to stabilise the reservoir as the injection is carried out. Make a 1 in incision at your site of choice and create a subcutaneous pouch large enough to take the reservoir.

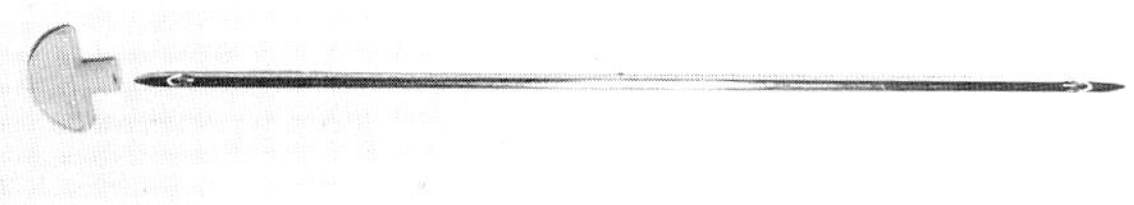

Figure 3. Tunnelling device.

- If you posses a tunnelling device (Figure 3), insert it from the central incision to the pouch; if the device is not long enough, you may have to do this in two stages via an intermediate stab incision in the loin. In this case tunnel from the central incision to the loin and then from the loin to the pouch. If you do not have a tunnelling device, then use your Tuohy needle in multiple stages with multiple stab incisions until you reach the pouch (Figure 4). Remember to use the Tuohy needle in the opposite way to a tunnelling device, i.e. from lateral to central, as otherwise you will not be able to remove the cannula!

  You will need to infiltrate widely with 0.5% Lignocaine; however, if the patient is well sedated, you will only need to infiltrate the skin at the entry and exit sites of the tunnelling device and the subcutaneous tunnelling itself will be well tolerated.

  When the tunnelling device is in position, you should pull out its introducer and feed the distal end of the catheter through it to emerge at the pouch or at the intermediate loin incision; if you have fed the catheter to the loin incision, you will need to reintroduce the tunnelling device and repeat the process until the catheter emerges at the pouch.
- When the epidural catheter has reached the pouch, attach it to the reservoir according to the manufacturer's instructions; the reservoir may then need to be anchored to the floor of the pouch, using a non-absorbable suture.
- Close the pouch and the central incision by suturing; the sutures can be removed after 5 days. Figure 4 shows a schematic representation of an implantable epidural device *in situ*.

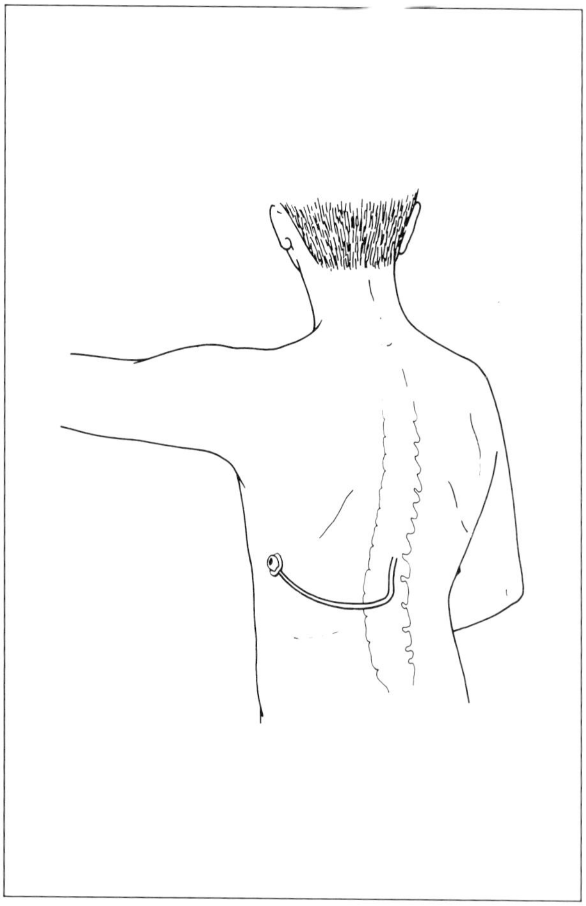

Figure 4. Schemating representation of indwelling epidural catheter device in situ.

Use of the Device

It may be advisable to inject 2% Lignocaine into the reservoir in order to achieve a rapid block; this will allow you to test the benefit of the system.

Most manufacturers recommend the use of Huber point needles, as these extend the life of the reservoir; in practice the authors use standard 25 gauge needles. It is vital that the patient or nursing advisors clean the area well before injection, since infection can often be a problem.

It is important to bear in mind the basic philosophy surrounding the use of an implantable device to deliver epidural opiates: your aim is to bind the opiates to receptors in the dorsal horn on a segmental basis. If the dose of opiate you select does not seem to be working, you probably need to increase the number of segments involved so that the opiate can affect more receptors. In this situation first try to increase the volume of dilutant (e.g. water) rather than the dose of opiate itself; increasing the dose of opiate to quasisystemic levels will inevitably bring in the opiate side-effects you are so keen to avoid.

We recommend that after you implant the catheter system you inject 2–5 mg of Diamorphine in 10 ml of water. We prefer diamorphine, as it comes in powder form free from preservative. It can be 'made up' as required; this obviates the necessity of keeping large stocks of preservative-free morphine. A not uncommon maintenance dose is 5 mg 12-hourly, bearing in mind our comments about segmental action.

You may encounter genuine tachyphylaxis after about 18 months to 2 years of continuous use. In this case rest the system from opiate for 24–48 h and use Bupivacaine instead. This seems to make the receptors opiate-sensitive once again.

You can, of course, use the catheter and reservoir system in conjunction with any of the commercially available programmable drug delivery pumps.

Note!
In our experience, the biggest problem in using these systems is infection. This may necessitate removing the system, controlling the infection and then reimplanting. In some cases, especially after repeated implants and in the presence of mild superficial infection, you can try resting the system completely, putting the patient on alternative analgesia and antibiotics. If the infection settles, you can then reuse the system; this, of course, requires careful judgement, since there is a direct conduit from the skin to the epidural space. Once the reservoir membrane starts to leak, fluid will accumulate in the tissues and this collection invariably becomes infected; the reservoir should therefore be changed for a new one.

Implantable Intrathecal System

A similar procedure is followed if you wish to implant a catheter into the subarachnoid space; the catheter in this case is obviously much finer but the principle is the same. A not uncommon use of this technique is for the infusion of intrathecal Baclofen in multiple sclerosis.

A useful rule of thumb is to give about one-tenth of the epidural dose by the subarachnoid route; thus, start with 0.2–0.5 mg of diamorphine. The volume of the carrying agent also needs to be much smaller, as the drug is distributed in the cerebrospinal fluid. We use about 1–2 ml of water.

There is always the risk of introducing infection and causing meningitis, so scrupulous aseptic technique is essential.

7

Stimulation Therapy

1 Spinal Cord Stimulation

Indications

- Ischaemic pain, particularly angina and peripheral vascular disease.
- Pain from spinal arachnoiditis.
- Phantom limb pain.
- Muscle spasm in multiple sclerosis.
- Any form of somatic pain not responsive to other techniques if funds to purchase the equipment are available.

Technique

The authors' experience has been gained using the Medtronic Pisces Quad system as well as the Neurotechnics Octrode system. The procedure is carried out in two stages. In the first stage a lead which carries the electrodes is inserted into the epidural space and is attached to a screener unit. This will give you an idea of the possible benefit of the system. If pain relief is achieved, the next stage is to attach the epidural lead to a permanent system. Here there are two choices:

1. Implanting a radiofrequency receiving antenna, used in conjunction with an external radiofrequency transmitter. In this system the patient controls both the 'on–off' function and the amplitude of the signal (Figure 1a–c). The external transmitter broadcasts its energy through the skin directly over the implanted passive receiver; this is connected to the lead in the epidural space by means of a special extension. The transmitter is connected by means of a cable to a special unit not unlike a TENS machine which the patient wears on his or her person. This system is of special use when a high output is required which would rapidly drain the battery of an internal system flat in a short time. External transmitters also provide a greater degree of flexibility in adjusting parameters after implant (Figure 2).
2. A totally implantable device (Implantable Pulse Generator – IPG). This device (currently in the UK, Medtronic only) has its own lithium batteries power source and control circuit, not unlike a cardiac pacemaker. The patient has control only over the 'on–off' function (Figure 3). With the totally internal system, the programming is done by the doctor, using a special console from the outside (Figure 4). Once the system has been programmed, the patient cannot change the parameters, but can only switch on and off with an external magnet.

With both systems, the patient can use the device during periods when the pain requires it. When the device is on, the patient will feel a slight tingling or buzzing sensation in the body parts that normally experience pain. The stimulation amplitude controls the level of perceived tingling which replaces the patients' pain.

Both systems can be implanted under either local or general anaesthesia.

Sedation

Sedation is not necessary for the first stage (insertion of temporary implant). It may be necessary for the second stage (insertion of permanent implant).

Antibiotic Cover

Antibiotic cover is advised: we suggest Ampicillin/Flucloxacillin, given as a loading intravenous dose of 500 mg of each with 250 mg of each orally 6-hourly for 5 days.

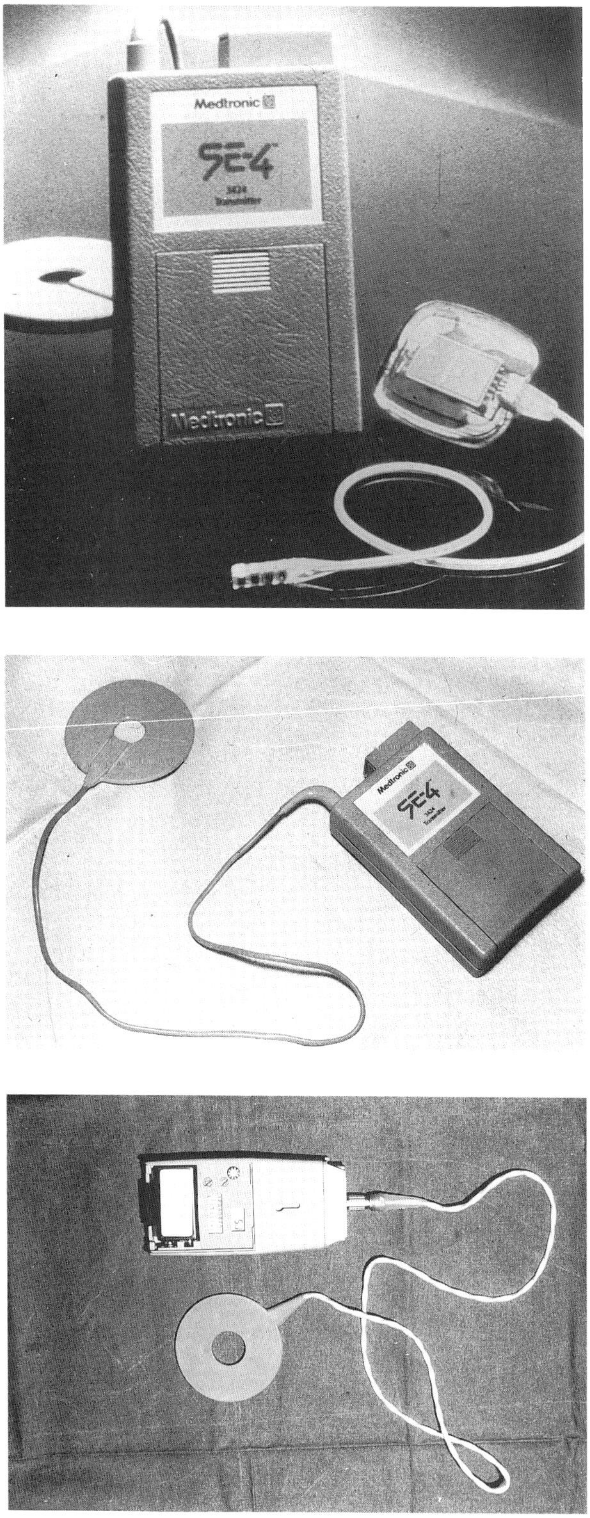

Figure 1.
a) Medronic RF stimulator and implantable receiving antenna.
b) Medronic RF stimulator and transmitting antenna.
c) Neurotechnic RF stimulator and transmitting antenna.

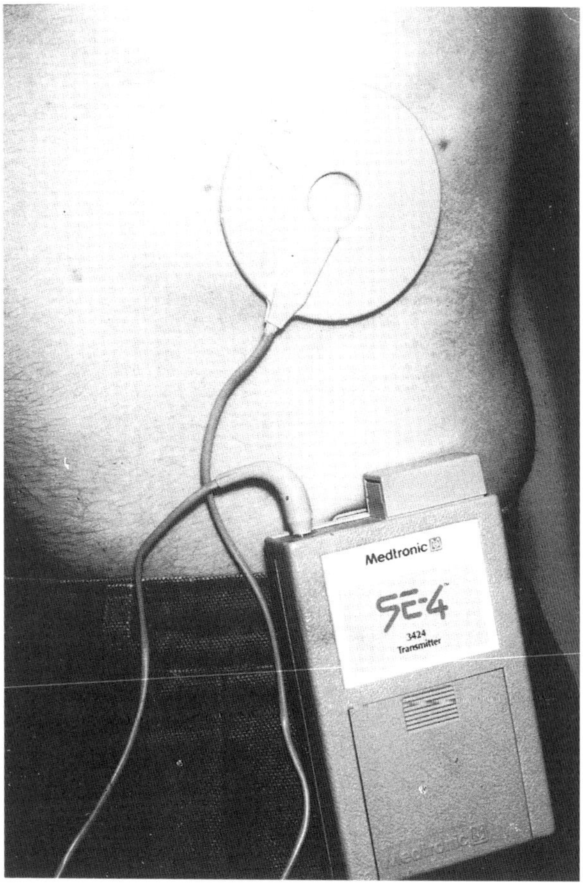

Figure 2. Patient wearing RF stimulator and transmitting antenna.

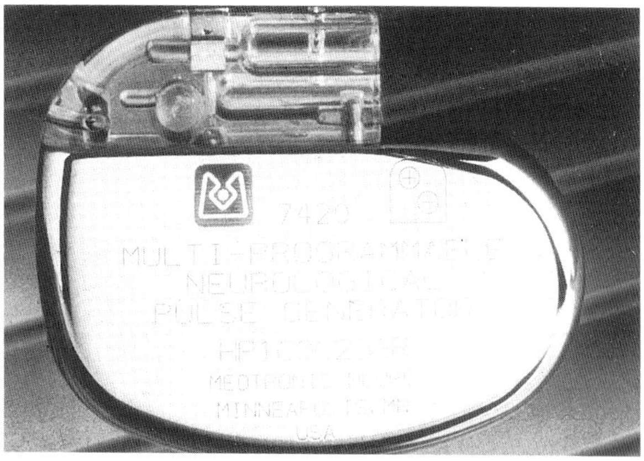

Figure 3. Implantable pulse generator.

Entry Points

- For pain originating in the lower extremities, you should insert your epidural needle at the L1/L2 intervertebral space or below; the lead should be

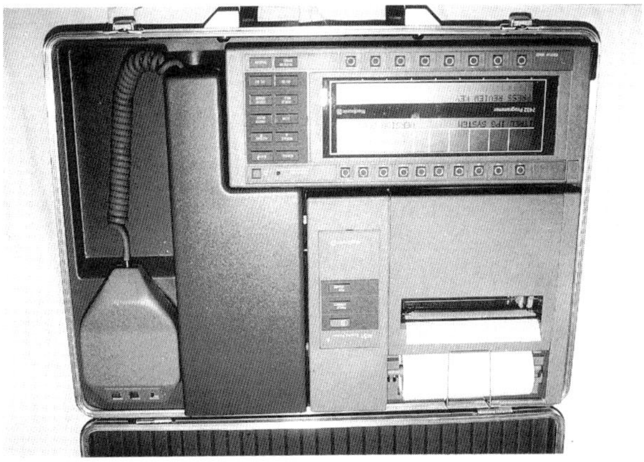

Figure 4. Physician's programmer.

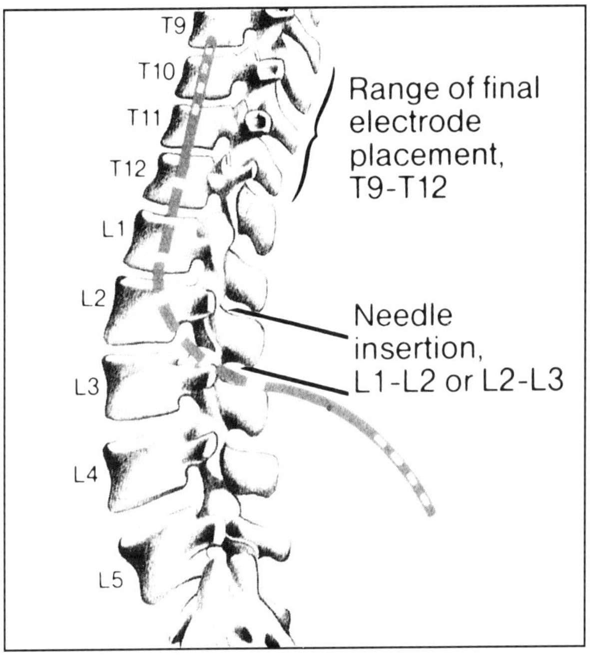

Figure 5. Lead placement for pain in the lower extremities.

threaded up to lie between T9 and T12 (Figure 5).

- For pain originating in the upper extremities, you should insert your epidural needle at the T3/T4 intervertebral space or below; the lead should be threaded up to lie between C4 and C5 (Figure 6a, b).
- For Anginal pain, you should insert your epidural needle at the T5/T6 intervertebral space or below; the lead should be threaded up to lie between C7 and T1.

Enough electrode lead should be inserted to allow fixation by the tissues; it is recommended that you allow at least three clear vertebral spaces from entry point to final placement site. The lead is rendered more stable if inserted obliquely through a paravertebral approach low down and manipulated to the desired higher level.

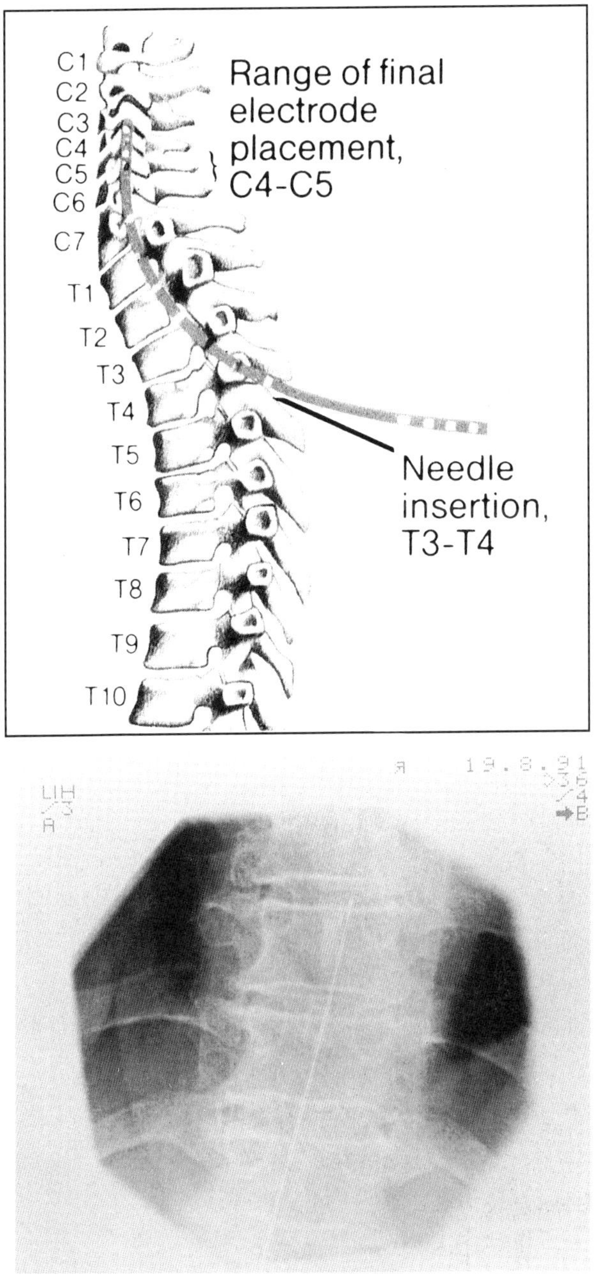

Figure 6.
a) Lead placement for pain in the upper extremities.
b) X-ray appearance of lead being threaded to the upper border of C7.

Position

The patient lies prone on a radiolucent table, with the C-arm image intensifier positioned for screening in the anteroposterior plane.

Use of the Medtronic System

Stage I: Insertion of Temporary Implant

a. The entry point is selected and marked after the exact level is confirmed by means of the image intensifier; the patient's back is then cleaned and draped.

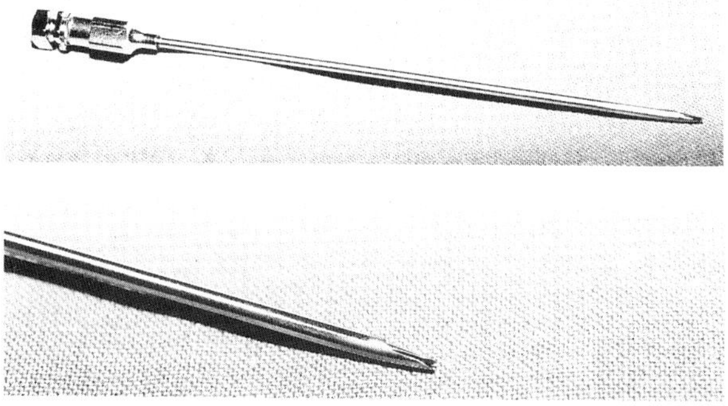

Figure 7.
a) Modified Tuohy needle.
b) Tip of modified Tuohy needle.

b. Insert the 15 gauge Tuohy needle with modified tip (Figure 7a, b) into the epidural space, using your preferred technique. As indicated above, a paravertebral approach allows an oblique entry into the epidural space, which makes the electrode easier to thread up to the desired level and prevents kinking. If you think you have entered the epidural space, confirm by passing the guide wire through the needle while screening with the image intensifier (Figure 8). This manoeuvre will also help create a track in the epidural space ready to accept the lead containing the electrodes (the 'lead').

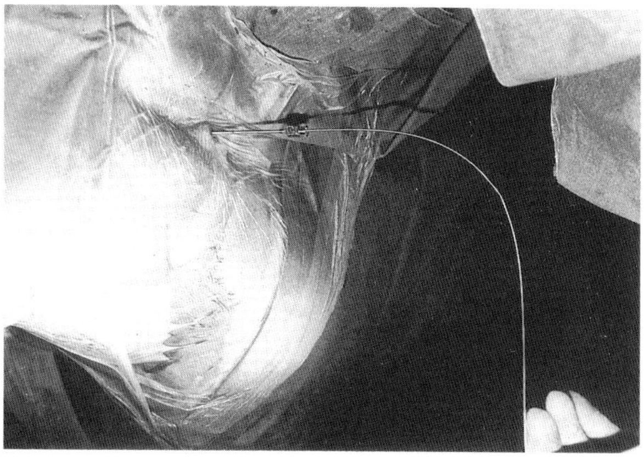

Figure 8. Guide wire being inserted through Tuohy needle.

c. When you are satisfied that you have entered the epidural space, take hold of the lead, which is mounted around a stiff stylet, and insert it through the Tuohy needle; the lead comes ready-mounted around a straight stylet for midline placement. You are also provided with a second stylet the distal end of which is curved for lateral placement, should this be indicated. This can be substituted for the straight stylet (Figure 9a, b). Repeated screening will reveal the direction which the electrode is taking (Figure 9c).

d. Advance the lead through to the desired vertebral level. The distal end of the lead contains four tiny electrodes numbered 0, 1, 2 and 3 from distal to proximal (Figure 10). The proximal end of the lead protrudes from the Tuohy needle around the stylet; the end of this stylet is attached to a plastic handle, which is now attached to the connector block on the temporary screening cable (Figure 11a).

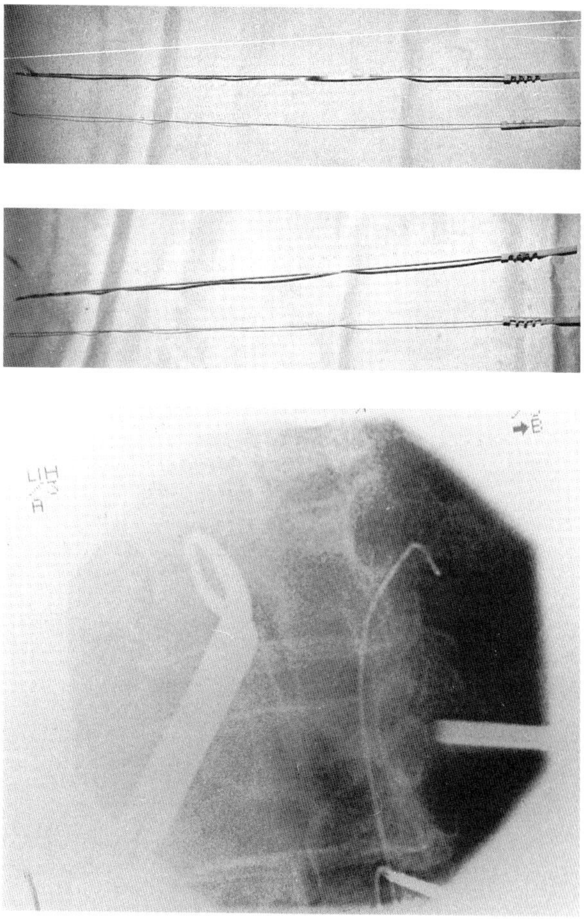

Figure 9.
a) Straight stylet inside electrode lead with curved stilette.
b) Curved stylet inside electrode lead with straight stilette.
c) Electrode emerging through nerve root foramen.

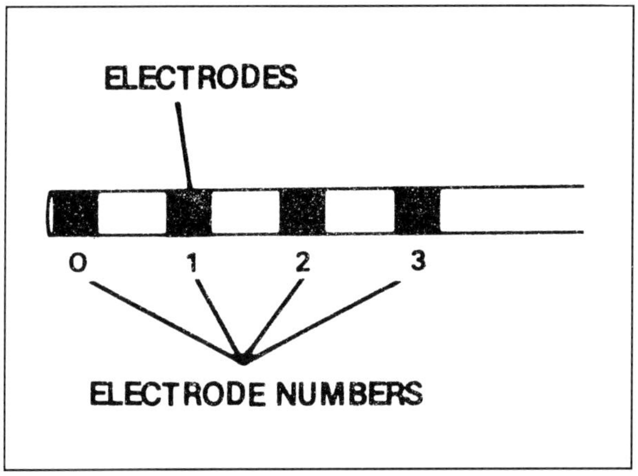

Figure 10. End of lead with four stimulation sites.

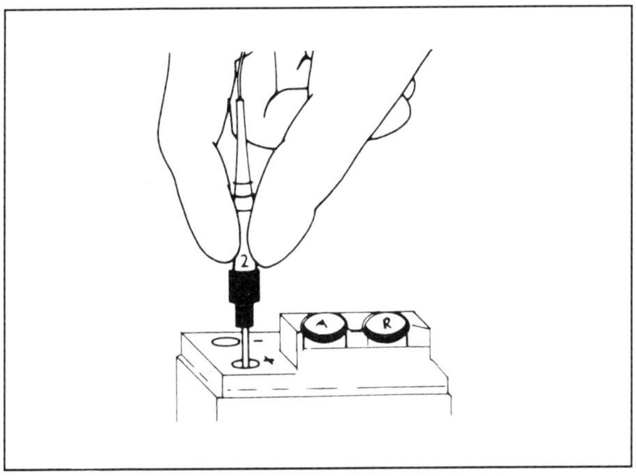

Figure 12. Connecting the jack plugs to the external screening stimulator.

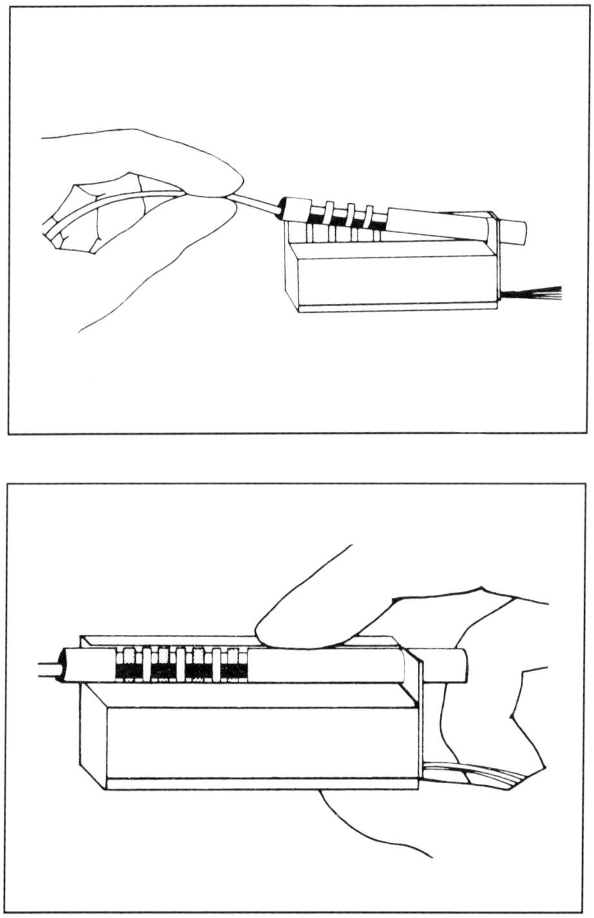

Figures 11a and 11b. Connection to external screening stimulator.

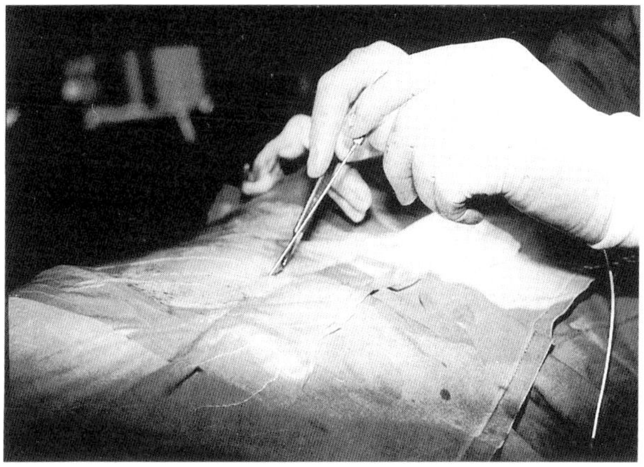

Figure 13. Cutting on to the Tuohy needle.

The connector block is provided with pins and onto these pins the plastic handle of the stylet should be aligned and pushed home (Figure 11b).

The screening cable ends in four jack-plugs each of which is numbered to coincide with the electrodes on the lead; these are used, two at a time, on the screening unit and you test various electrode combinations and parameters – i.e. rate, amplitude and pulse width – in order to achieve paraesthesiae at the target site (Figure 12). Manipulation of the lead to different vertebral levels may be necessary.

e. When you are satisfied with the final position of the lead, turn the screener off and disconnect the connector block from the stylet handle; cut down around the Tuohy needle until you reach tough fascia (Figure 13); remove the stylet from the lead, being careful not to disturb the position of the lead (Figure 14); at the end of this manoeuvre, in order to ensure that the lead tip has not in fact moved, you can restimulate by pressing the lead contacts against the appropriate pins of the connector block.

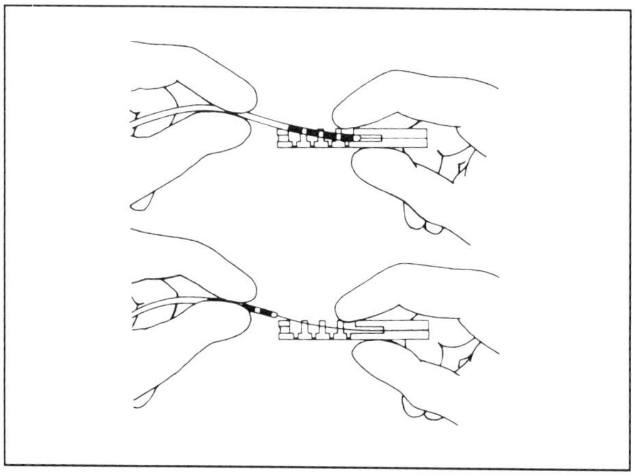

Figure 14. Removing the stilette from the lead.

f. Remove the Tuohy needle, taking great care not to pull out the lead; proceed in the same way as you would in removing any Tuohy needle containing an epidural catheter. The bevel of the needle is constructed to prevent shearing of the lead as the needle is gently pulled out of the epidural space. Slide the anchoring device over the end of the lead and pass a suture around the device and through its eyelet to anchor firmly into the deep fascia (Figure 15 a–c). Suturing the anchoring device to the deep fascia can be difficult and frustrating to the non-surgeon! In recent year the authors have adopted David Cherry's technique of inserting the Tuohy needle through a deep paravertebral incision which allows the silk stay sutures to be attached to the fascia before the needle is inserted. They are then already in place at this stage and merely need to be tied firmly around the anchoring device.

g. You now have to plan the skin exit site for your percutaneous extension, at which site you must make a small stab incision; this had best be on the side of the dominant hand, so that the permanent device can be implanted on the opposite (non-dominant) side. Take the percutaneous extension and place one end in contact with the end of the lead and move the other end laterally to the side of the body; this will show you where to place your stab incision.

h. Now turn your attention back to the incision around the Tuohy needle. Burrow laterally with the Tuohy needle inserted through the central incision to the point where you wish to create a small pouch which will hold the junction of the percutaneous extension and the lead (Figure 16).

i. Take the tunnelling device (Figure 17a) and remove the 'screw-on' sharp end (Figure 17b). Hold the handle end of the device at the stab incision and the other end at the central incision, and bend it so that it moulds

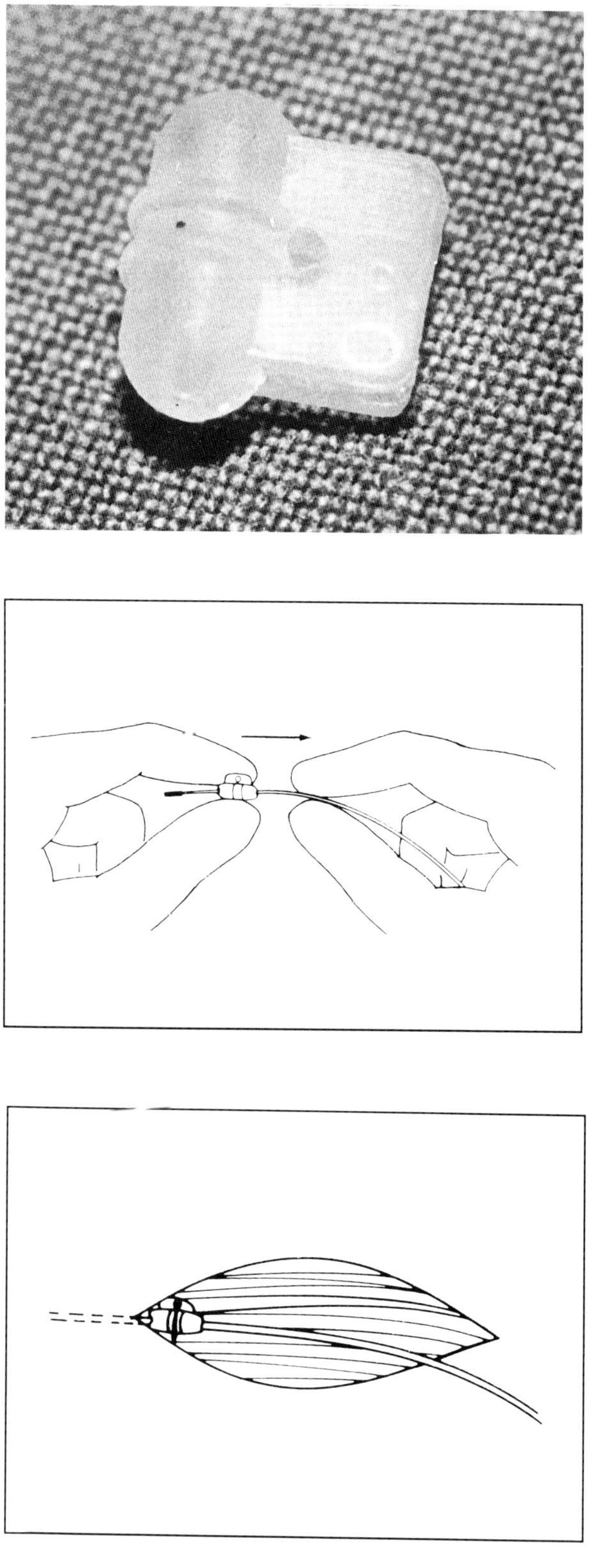

Figure 15.
a) Anchoring device.
b) Anchoring device being slid over lead.
c) Anchoring device sutured in place.

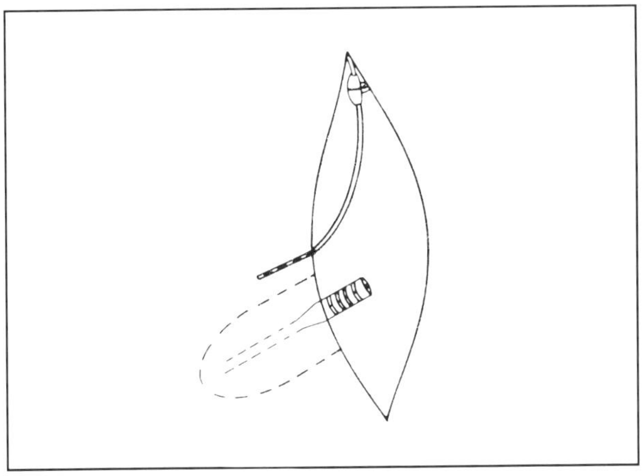

Figure 16. Proximal connecting of percutaneous extension and siting of possible percutaneous pouch.

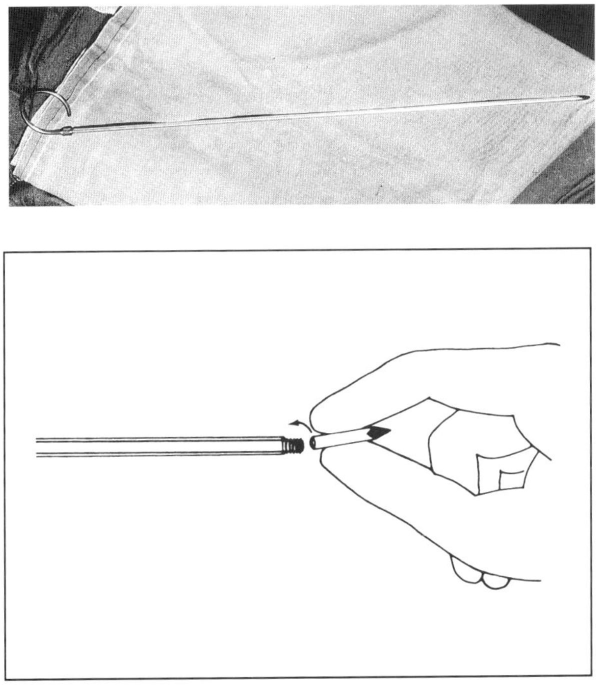

Figure 17.
a) The tunnelling device.
b) Removing the sharp tip of the tunnelling device.

to the contour of the body. You can now make good use of the Teflon tubes which are used to package the lead and the percutaneous extension. Slide the Teflon tube over the device; replace the sharp 'screw-on' end and push the device from the stab incision so that it emerges at the lateral pouch you have just created. Remove the sharp end again and pull the device out from the lateral incision, leaving the Teflon tube *in situ* (Figure 18).

j. Take the percutaneous extension cable and insert

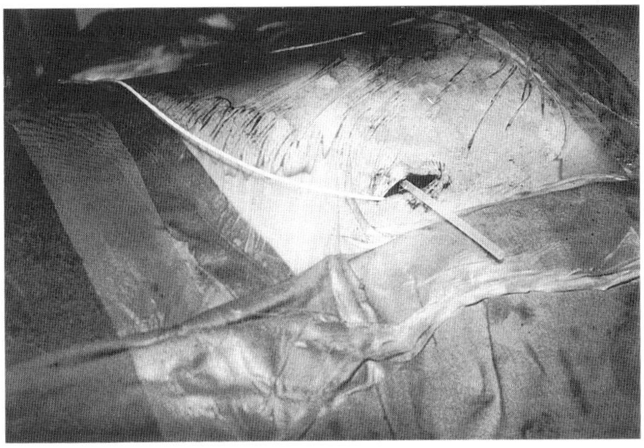

Figure 18. Teflon tube emerging through the central incision.

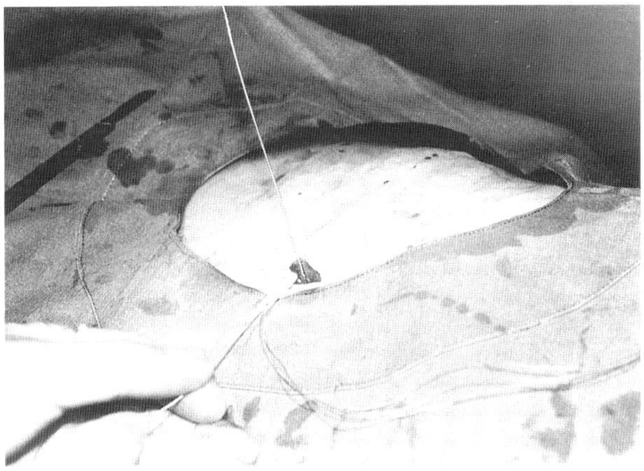

Figure 19. Percutaneous extension being inserted in to Teflon tube using guide wire.

its pin connection into the Teflon tube at its medial end (Figure 19). Using the guide wire, push it into the tube so that it emerges at its lateral (stab incision) end (Figure 20). When this has been done, remove the Teflon tube, leaving the cable *in situ*. Now fit the boot over the end of the lead protruding from the lateral pouch (Figure 21). Take the end of the lead and push it firmly into the socket of the setscrew connector junction end of the percutaneous extension (Figure 22); observing the junction while gently rotating the lead will help to determine whether the lead is properly inserted.

Once the lead is completely inserted, tighten each of the four setscrews by turning clockwise in the setscrew sockets with the hex wrench provided; avoid excessive tightening (Figure 23). Slip the boot over the connection thus created, tie both ends of the boot to grip both lead and extension firmly (Figure 24), and place this connection carefully in the small subcutaneous pouch previously created.

k. Attach the pin connection of the percutaneous extension cable to the screening cable connector block as you did earlier for the lead (refer to Figure 11); leave the temporary device in for a minimum period of 1 week and a maximum period of 3 weeks, to get a good idea of the benefit or otherwise of the system.

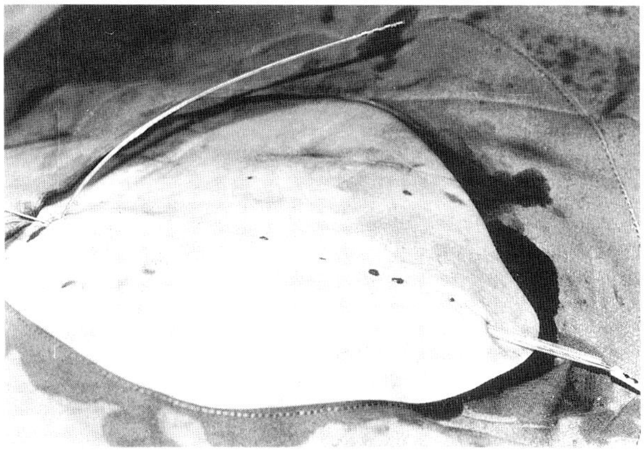

Figure 20. Percutaneous extension emerging (right) from stab wound and off Teflon tube.

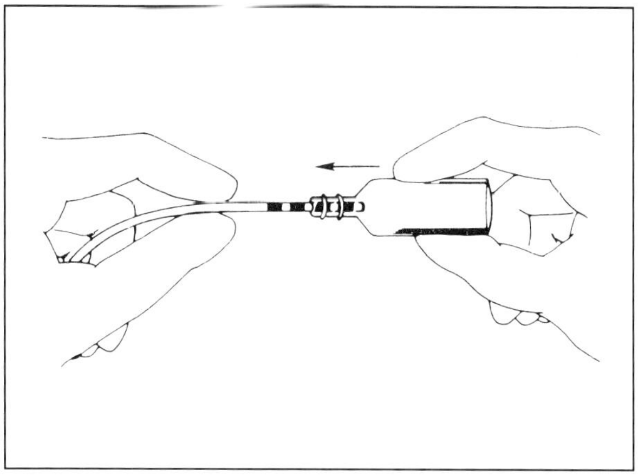

Figure 21. Boot being put over distal end or lead.

Stage I (a)

If there is *no benefit* from the system, you will have to remove the lead, as follows.

- Expose the lead percutaneous extension junction, i.e. open up the central incision and lateral pouch and deliver the junction from the lateral pouch.
- Cut through the anchoring device, freeing the lead from the surrounding fascia.
- Ask your assistant to take hold of the screening cable

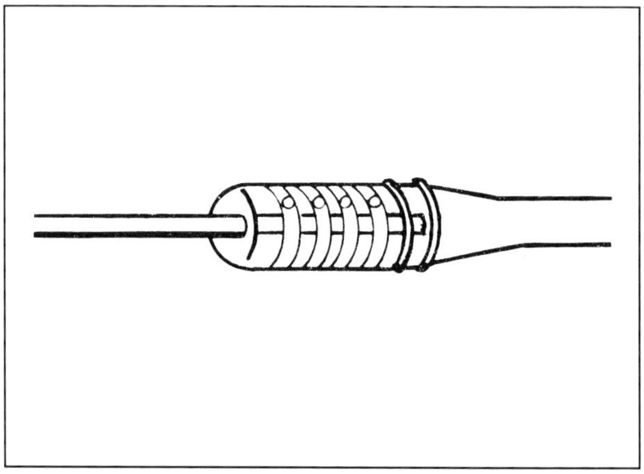

Figure 22. Distal end of the lead inserted in to the medial end of the percutaneous extension.

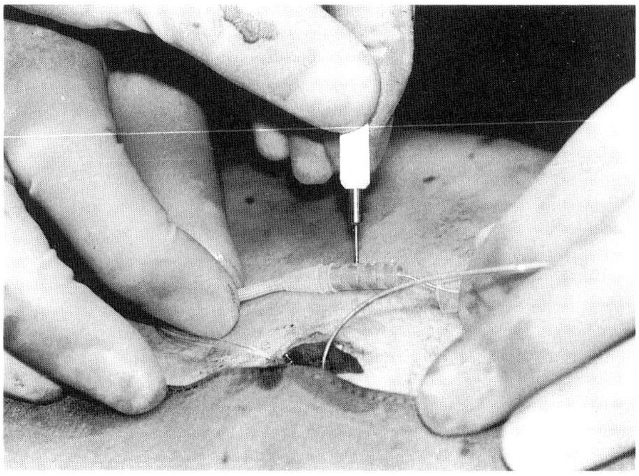

Figure 23. Extension being tightened on to lead.

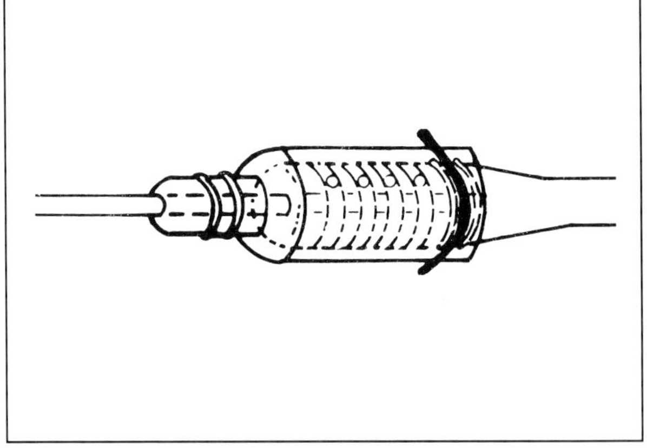

Figure 24. Fitting the boot.

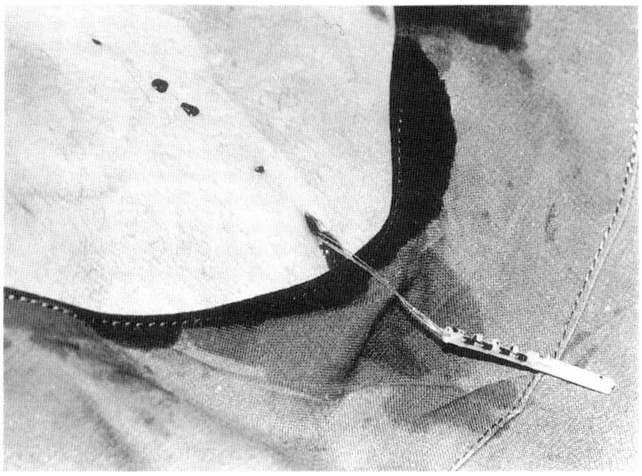

Figure 25. Screening cable connector, block pin connection emerging from distal stab incision.

connector block-pin connection and to pull gently, thus exposing a length of sterile wire as it exits through the stab incision (Figure 25); cut this sterile wire with a sterile pair of scissors. Take hold of the lead–percutaneous extension junction and pull so that the distal. end of the extension wire is delivered through the lateral pouch (sterile wire traverses sterile tunnel).

- Pull the lead out of the back and suture the lateral pouch and central incision.

Stage II

If the system is *beneficial*, you will need to implant a permanent electrode lead as follows.

a. Expose the lead–percutaneous extension junction, i.e. open up the lateral pouch and deliver the junction from it.

b. Cut through the sutures holding the boot snugly round the lead and extension wire; pull the boot back over the lead. You can cut the boot open if you want to, but take great care not to damage the lead itself as you cut.

c. Loosen the four setscrews, using the hex wrench, and remove the lead from the percutaneous extension. Remove the boot off the lead. Cut the percutaneous extension wire just distal to its former junction with the lead and pull it out through the stab end. You now have a free lead in the lateral pouch still firmly anchored to the deep fascia by the anchoring device. Open the central incision and pull the lead back from the lateral pouch so that it now emerges from the central incision. Create a new lateral pouch on the opposite (non-dominant) side, using a Tuohy needle as before (see h above, page 70).

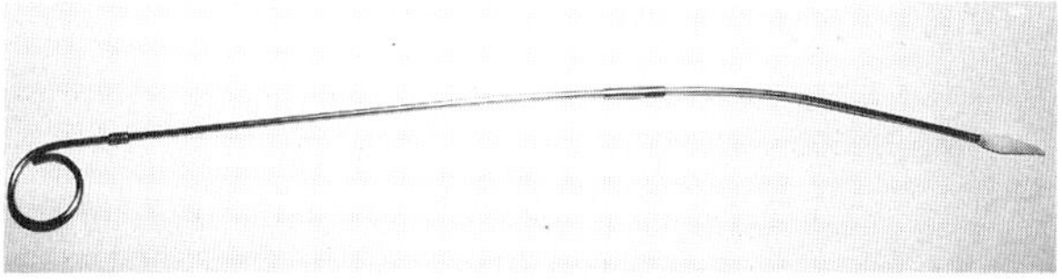

Figure 26. Tunnelling device with arrow head.

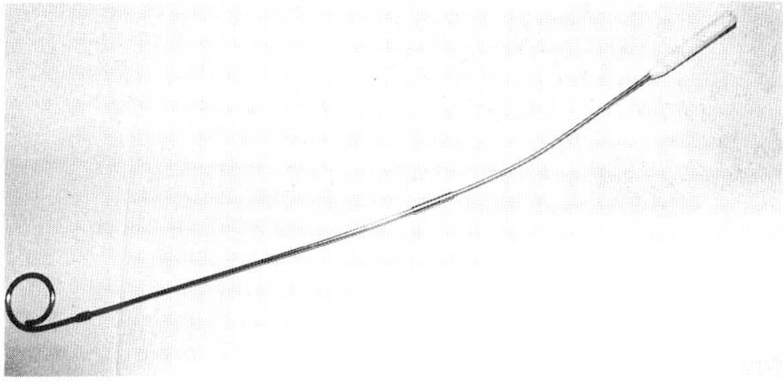

Figure 27. Tunnelling device with extension carrier.

d. You now need to decide the location (on the non-dominant hand side) of the pouch into which will be embedded either the radiofrequency receiver or the Itrel IPG (implantable programmable pulse generator); to do this you will need to take the tunnelling device, which in the 'permanent extension kit' is in two parts. Assemble and mould the tunnelling device over the side of the body from the centre to the side, so as to shape it to the body contour. Make a 3 in incision in the abdominal wall and create a pouch, taking great care to ensure that the site of the pouch is not too close to the rib cage or it will be very uncomfortable when the patient bends forward.

e. Attach the 'arrowhead' to the end of the tunnelling device (Figure 26) and push it through from lateral pouch to abdominal pouch so that the arrowhead is through the skin. Unscrew the arrowhead from the tunnelling device and, in its place, fix the plastic 'extension carriers' (Figure 27).

f. Into the extension carrier fix the end of the radiofrequency antenna lead or the IPG extension lead (Figure 28) and then pull the tunnelling device out through the lateral pouch so that the medial end of the

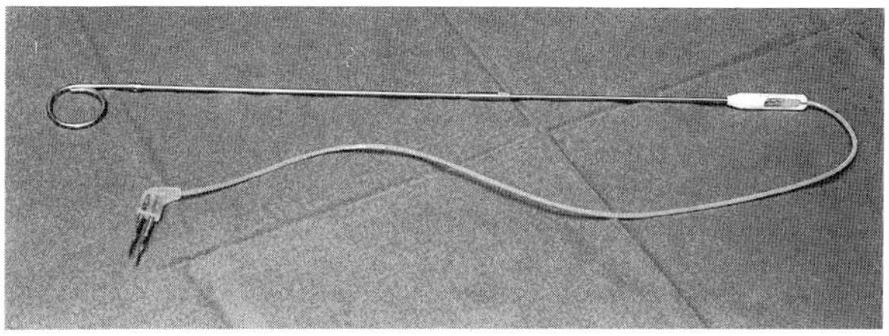

Figure 28. Tunnelling device with carrier.

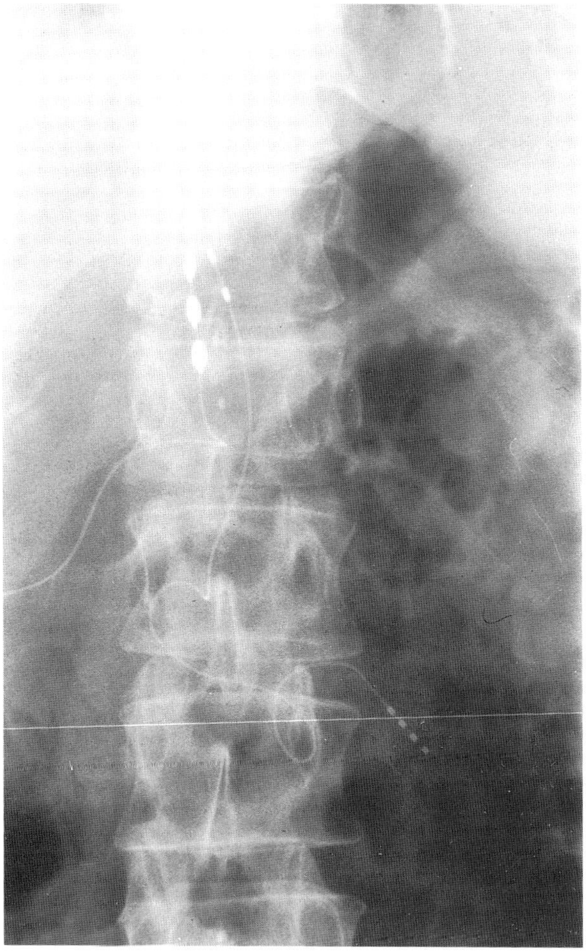

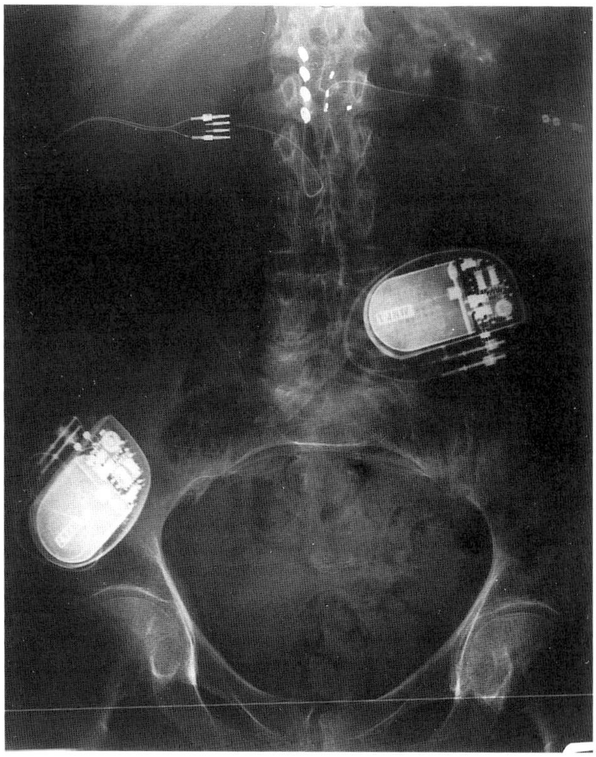

Figure 29.
a) Patient with two spinal cord stimulator leads Resume and Quad.
b) Patient with two spinal cord stimulator leads Resume and Quad showing IPG's

radiofrequency antenna lead or the IPG extension lead comes out through the lateral pouch.

Put the protective boot over the epidural lead in the same way as you did for the temporary device. Now insert the epidural lead fully into the medial end of the radiofrequency antenna lead/IPG extension lead and tighten the screws. Pull the boot over and suture exactly as before, and secure the boot at both ends with non-absorbable sutures as already described.

The extension lead must now be attached to the IPG or RF receiver; the end of the extension lead is pushed into its socket on the IPG or RF antenna. Secure the extension pins by tightening the setscrews.

g. Ensure that the pouch for the radiofrequency antenna or the IPG is large enough to accommodate the device but not so large that the device is able to move about in its pouch. Secure the IPG through with a suture through its available slot. Make sure that there are no sharp bends in the lead; any excess lead should be collected into a circle.

Suture all your incisions.

Programming

The IPG settings are operated via an external console and the radiofrequency antenna via an external transmitter; we refer you to the manufacturer's booklet for this.

Medtronic Resume Lead

This is a wider electrode which is placed by the surgeon at laminectomy in the epidural space under direct vision. The indication for this lead is when obliteration of the epidural space makes it impossible to place the lead percutaneously–e.g. in the presence of gross arachnoiditis and scarring or when a long-term spinal cord stimulation lead placed on one side of the cord has scarred up the space, making the percutaneous placing on the other side impossible. Thus, you may have placed a lead percutaneously into the right side of the epidural space for right lower limb ischaemic pain; the patient may after a long interval develop ischaemic pain in the left lower limb but percutaneous insertion of a lead into the left side is impossible, as the epidural space has become obliterated by scarring. In this situation direct placing of the lead in the epidural space at open laminectomy may be the only solution (Figure 29a, b).

The Neurotechnic Octrode System

The essential difference between the Neurotechnic Octrode and the Medtronic systems is that the former has eight stimulation points at the electrode lead tip as opposed to the four on the Medtronic (Fig. 30). In the Octrode, the electrode lead is not mounted on a stylet as is the Medtronic. With the Octrode, you are provided with a floppy guide wire which is mounted on a rigid stylet and this is introduced through the Tuohy needle so as to create a track for the electrode lead to follow (Fig. 31).

If the stylet become twisted as you attempt introduction, you simply cut off the twisted part, together with an equal length of floppy guide wire to recreate a straight system.

When the guide wire reaches the desired level, remove it and quickly insert the Octrode lead through the track created for it. The other important difference between

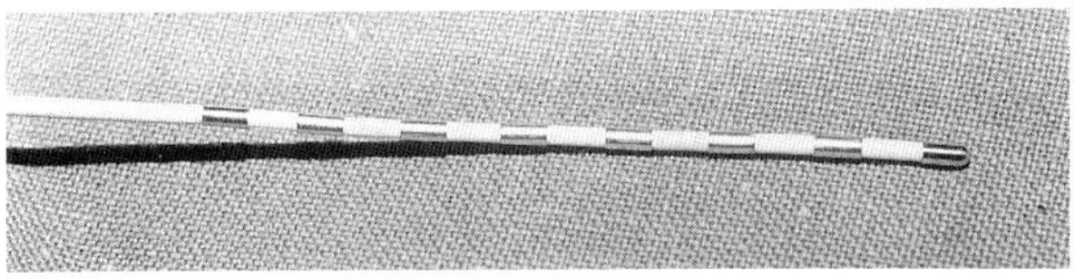

Figure 30. The Octrode lead.

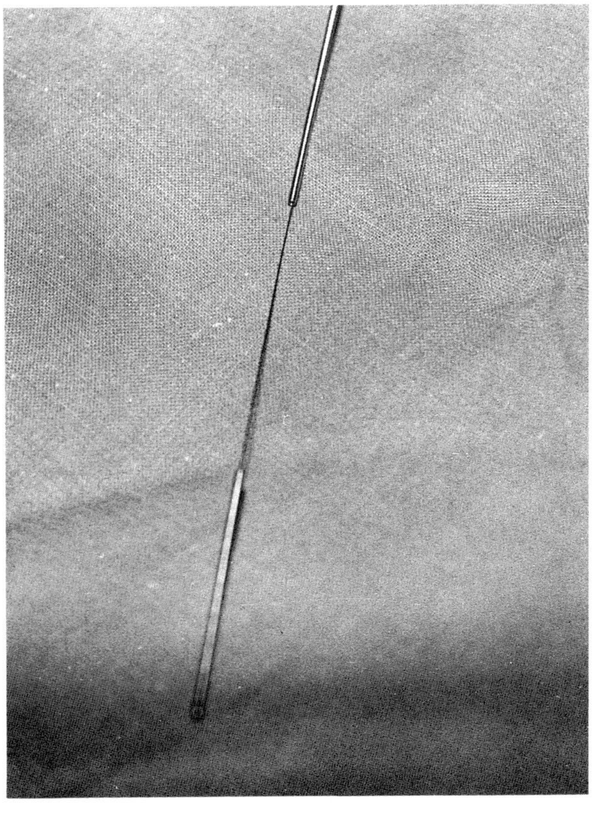

Figure 31. Floppy guide wire with rigid stylet.

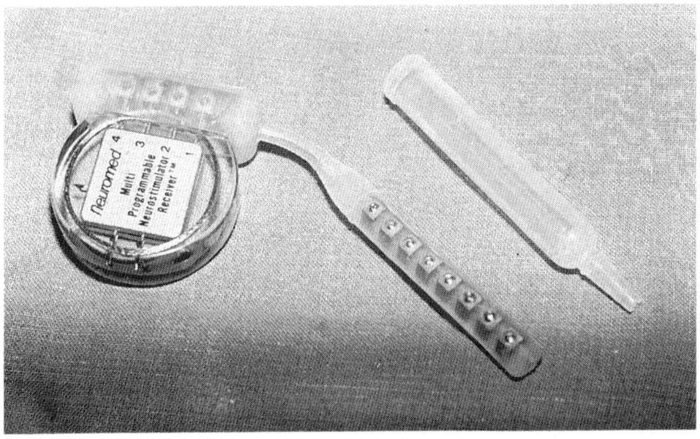

Figure 32. Octrode RF receiver.

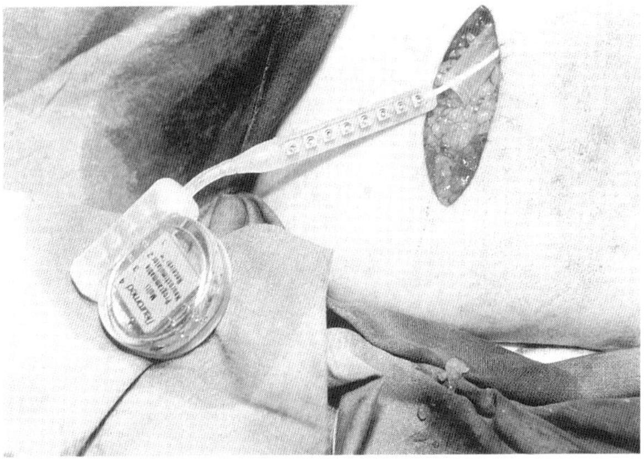

Figure 33. Octrode lead attached to RF receiver.

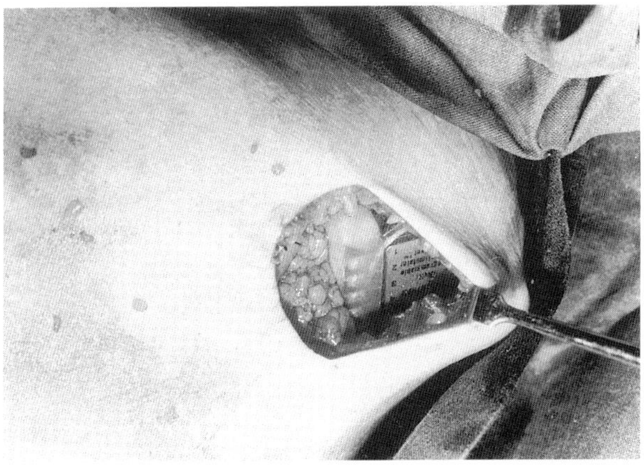

Figure 34. Octrode receiver in subcutaneous pocket.

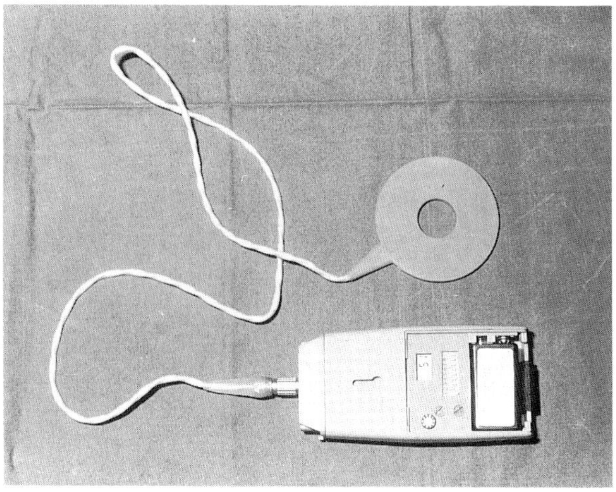

Figure 35. Octrode stimulator and antenna.

the two systems is that the Octrode lead is connected to the external screener by screw connections and not via a connecting block. The Octrode as presently available in the UK is purely a RF system.

The tunnelling process is identical with that of the Medtronic save that the sharp end does not screw off; the Teflon tube fits over the tunnelling device without the need to remove the sharp end.

In STAGE 2, with the Octrode, there is no intermediate connection, the end of the lead fits directly into the RF system (Figs. 32, 33 and 34).

Figure 35 shows the RF antenna and stimulator for the Octrode.

2 Transcutaneous Electrical Nerve Stimulation (TENS)

TENS can be used in any type of pain, although central pain seldom responds. It is important to ensure that the patient is physically and mentally able to handle the TENS unit.

Wave Form

Most TENS units produce a square wave of variable amplitude and pulse width. The wave is delivered in a monophasic form, i.e. by direct current. The current flows from the negative to the positive electrode.

Variable Parameters

These are:

- Rate, i.e. the frequency in Hertz or pulses per seconds.
- Pulse width, i.e. the duration of the wave delivered by the unit.
- Amplitude, i.e. the strength of the current in milliamps, which determines the height of the wave.
- Duration of stimulation.

Modes of Delivery of TENS

The modes of delivery of TENS are (a) conventional and (b) acupuncture-like.

Conventional TENS (High-Frequency Low-Amplitude)

Here the aim is to stimulate principally the mechanoreceptor A-beta fibres and to 'close the gate'.

- Rate: 80–100 Hz.
- Pulse width: 100 μs.
- Amplitude: go for the maximum amplitude which the patient can comfortably tolerate producing a definite tingling sensation in the area of the pain.
- Duration of stimulation: as a minimum go for about 8 h/day; however, many patients choose continuous stimulation. It is now recognised that while the TENS unit is in use, the neurotransport mechanisms are 'turned off'; thus, neuropeptides cannot be transmitted centrally. This means that continuous use allows the levels of neuropeptides in the dorsal horn to fall and the pain threshold to rise back to normal. Thus, the longer the TENS unit is used, the better chance there is of obtaining a stepwise permanent improvement.

Acupuncture-like TENS (Low-Frequency High-Amplitude)

The aim here is to stimulate also deeper-seated afferent A-delta fibres apart from the mechanosensitive fibres; it may work where conventional TENS has failed, e.g. when the pathological process has disrupted the standard afferent pathway. It may be the form of choice in neuropathic pain with dysaesthesia.

The main feature is that the impulses are delivered in short bursts, usually a train of 6–8 stimuli delivered at 1–2 Hz.

- Rate: train of seven pulses delivered at 1–2 Hz.
- Pulse width: usually greater than 100 μs.
- Amplitude: go for the maximum amplitude which the patient can comfortably tolerate and which produces definite muscle twitches; most workers prefer to elicit sensory stimulation in the area, not muscular twitches. This latter, although not really

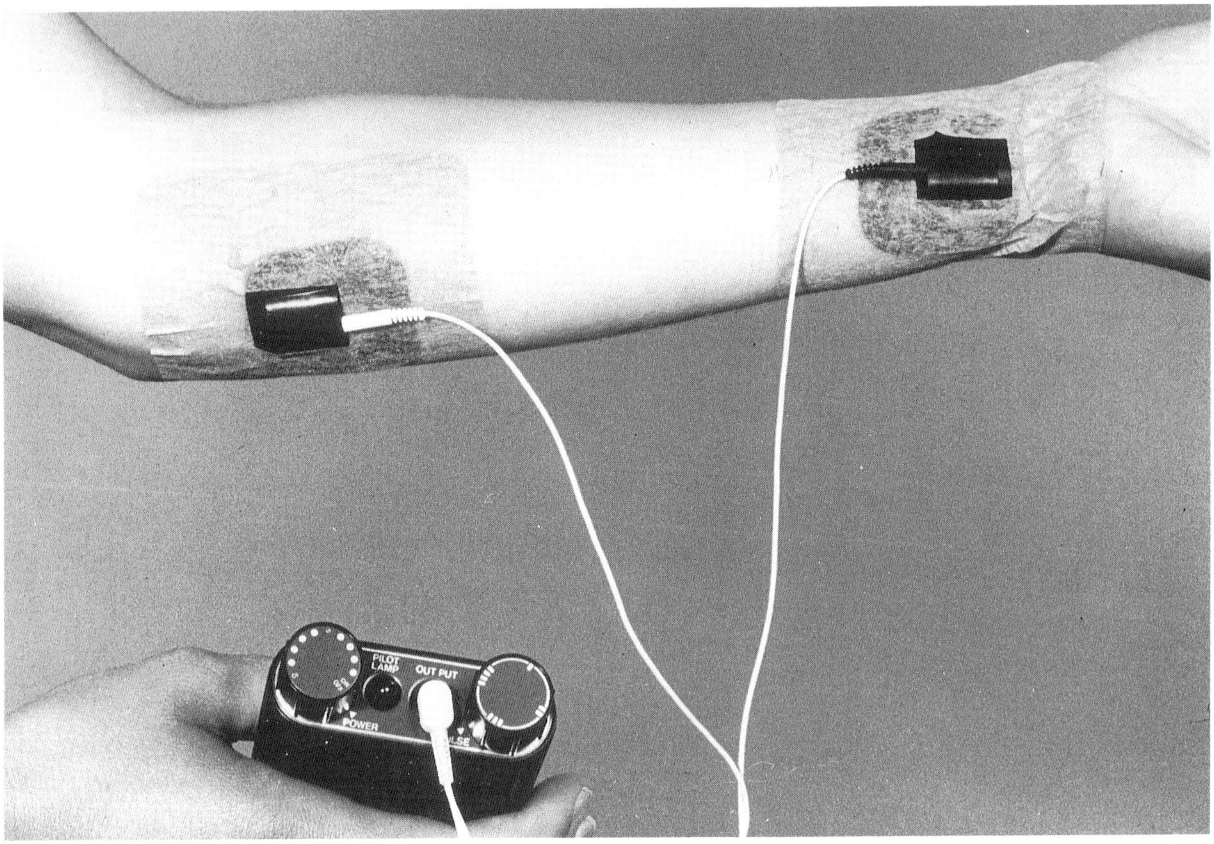

Figure 1. TENS unit in situ.

acupuncture-like TENS, may be more comfortable for the patient and is usually referred to as the 'burst' mode of stimulation. This form of TENS usually employs a higher amplitude than does conventional TENS.

- Duration of stimulation: usually much less than that of conventional TENS. When benefit is obtained, it usually manifests itself after a shorter period of stimulation than with conventional TENS. In general, aim for 20 min of stimulation, 2–4 times per day; however, as with conventional TENS, some patients may choose continuous stimulation.

Placement Site for Electrodes

Conventional TENS: Trigger spots, dermatomes and peripheral nerves supplying the painful area. Electrodes may not be tolerated over the painful area itself if hyperaesthesia is present. There is no point in placing the electrodes over numb areas, since, for TENS to work, sensation must be present.

Acupuncture-like TENS: over nerves innervating related myotomes; also over motor points.

TENS can be used bilaterally for reinforcement. In some cases of hyperaesthesia you may obtain benefit by placing the electrode on the opposite side at the same segmental level.

Care of Electrodes

If you are using gel in order to obtain electrical contact between the electrodes and the skin, clean skin and electrodes every 24 h and apply fresh gel, as skin irritation can develop. Should the patient develop an allergy to the electrode gel, you can try mixing a small amount of Hydrocortisone cream in with the gel. Disposable pregummed hypoallergic electrodes are available for patients who develop allergies. Karaya Gum electrodes are also useful, as they are the least allergenic, but patients do not seem to like using them, as they quickly become sticky and messy. In some cases the patient develops an allergy to the paper tape used to hold the non-pregummed electrodes in place. In these cases a very handy commercially available product it 'Skin Prep', which is used for colostomies.

Caution!
Be careful if your patient has a cardiac pacemaker *in situ*, as TENS equipment can interfere with its function; check with the manufacturers.

8

Regional Blocks

1 Neuro Modulation

It is not always a good idea to proceed to neurolytic block in those patients with chronic pain from a non-malignant cause in whom a local anaesthetic block has given good, albeit short-acting, relief. In these cases you may actually create neuropathic or deafferentation pain. These patients have a normal life expectancy, so they will not thank you for this! Here, then, you should try to elicit long-term relief, not by creating a neuronal discontinuity but by selective interference with nerve function.

The three principal techniques employed by the authors are as follows.

1. *Injection of a mixture of steroid plus Bupivacaine.* The steroid probably acts by its general membrane-stabilising properties, i.e. it reduces neuronal activity in a non-specific manner. The effect may be quite long-acting, e.g. as long as 3 months from a single injection. There are various depot steroid preparations commercially available: the most popular are Methylprednisolone Acetate and Hydrocortisone Acetate. Since these preparations contain preservatives, some authorities suggest that they should not be used in the vicinity of the spinal cord and they recommend instead Hydrocortisone Succinate, a non-depot preparation.

2. *Injection of streptomycin.* This antibiotic seems to affect cell membrane function and it may sometimes produce pain relief when given by local injection (1). In view of the risk of ototoxicity we always carry out an audiogram and caloric tests on the patient and proceed with therapy only if these are normal; we carry out nerve blocks, using 1 g of streptomycin in 2–5 ml of 0.5% Bupivacaine on a weekly basis for 5 weeks. There is always a risk of a local hypersensitivity reaction to streptomycin; this can be severe enough to mimic the effects of a localised abscess. For this reason we would caution against injecting streptomycin around spinal nerve roots.

3. *Guanethidine.* Guanethidine acts by depleting the sympathetic nerve endings from noradrenaline; it is, of course, used in the procedure of intravenous Sympathetic block (see page 53). We have also used it in lumbar and stellate ganglion blocks, where we mix 5 mg of Guanethidine with the Bupivacaine. This can enhance the duration of a straightforward local anaesthetic sympathetic block.

Guanethidine may also be of use in perineal pain when one can carry out bilateral pudendal nerve blocks with Guanethidine and Bupivacaine (Foster, J.M.G., personal communication).

2 Cervical Nerve Root Injection

Indication

Cervical root pain.

Clinical Picture

Pain along the distribution of any one or more of the cervical nerve roots. Very often X-ray examination reveals intervertebral foraminal encroachment.

Technique

Sedation

Sedation is often necessary.

Position

The patient lies supine with the head on a radiolucent rest which affords full access to the neck; a C-arm image intensifier is positioned as shown in Figure 1.

Landmarks

Identify the mastoid process and mark its apex; draw a line from here down along the posterior border of the sternomastoid. Two finger's breadth below the apex of the mastoid process corresponds approximately with the entry point for the C2 nerve root (Figure 2); subsequent nerve roots are approximately one further finger's breadth below the point above.

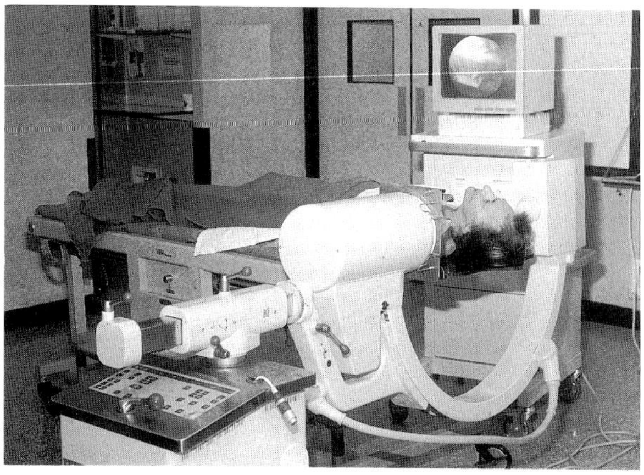

Figure 1. Position of patient for cervical nerve root injection.

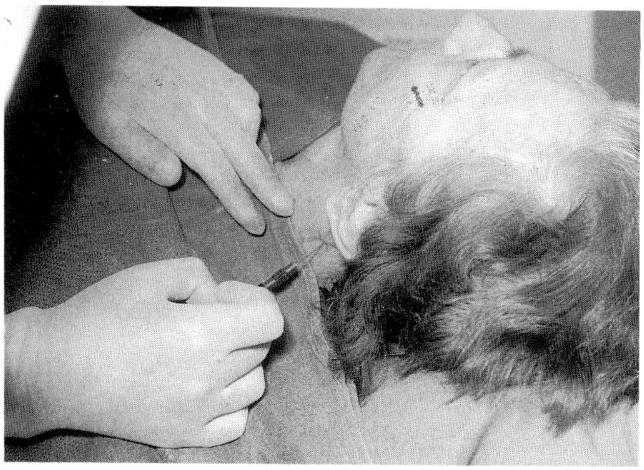

Figure 2. Entry point for C2 nerve root.

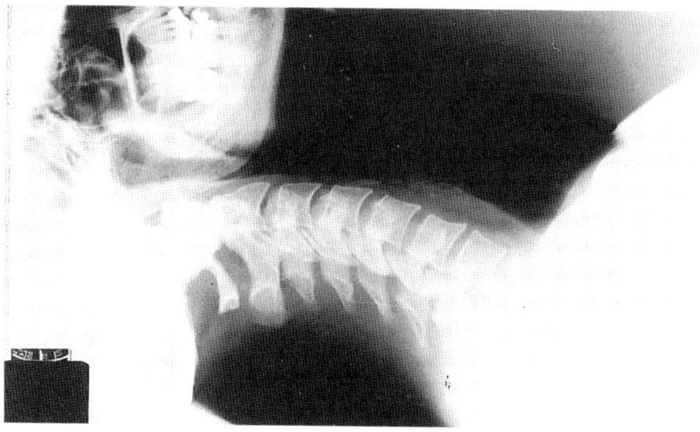

Figure 3. Lateral view of cervical spine with image intensifier.

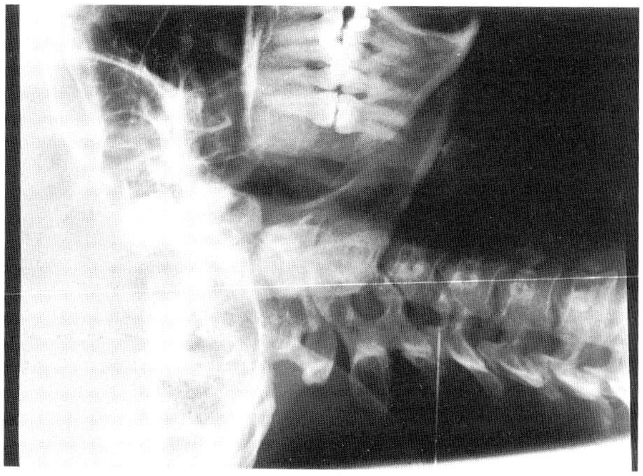

Figure 4. Oblique view of cervical spine with image intensifier. Note needle tip at posterior margin of intervertebral foramen.

View the cervical spine with the image intensifier positioned as shown in figure 1(b); this gives you a lateral view of the spine (Figure 3). Rotate the intensifier C arm so as to obtain an oblique view of the spine; in this view, the intervertebral foramina and their associated facet joints are clearly seen (Figure 4).

Clean and drape the area.

Procedure

- Infiltrate the entry point with 0.5% Lignocaine with Adrenaline.
- Insert a 21 gauge 5 cm (2 in) needle through the skin and direct it towards the intervertebral foramen in question. The needle should come to rest on bone at the posterior (dorsal) margin of the foramen (Figure 4) and then be 'walked' just into the posterior part of the foramen.

  In the case of the C2 nerve root, there is no foramen as such; here the needle should be positioned as it

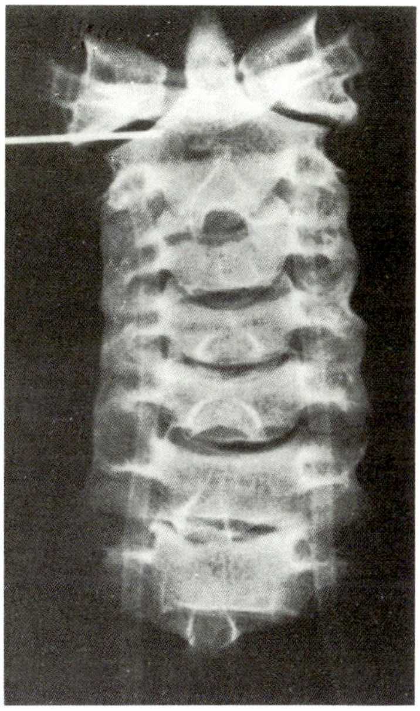

Figure 5. Medial tip of needle in cervical nerve root injection.

would be for carrying out a percutaneous cordotomy (see page 30).

Rotate the image intensifier so as to screen in the anteroposterior plane and confirm that you have advanced the needle no further medially than an imaginary line passing through the cervical facet joints (Figure 5).

- If you are carrying out a *diagnostic* injection, remove the thermocouple probe from the needle and inject either 1 ml of 2% Lignocaine or 0.5% Bupivacaine, depending on whether you want an immediate short-term or a prolonged effect. Naturally, with a pole needle you can carry out a direct injection through the needle.
- If you are carrying out a series of *therapeutic* injections, when you are satisfied with the position of your needle, move on to the next root. We recommend that no more than three cervical nerve roots be injected at one sitting. When all the needles are correctly positioned, inject 1 ml of 0.5% Bupivacaine together with your steroid preparation through each.

The reason it is best to carry out all the injections when all the needles are in place is that the local anaesthetic injected at one level may trickle onto the adjacent root and interfere with your subsequent stimulation manoeuvres.

After-care

It is prudent to keep the patient in hospital for at least 1 h after the procedure; observe in case you have accidentally injected Lignocaine into the CSF!

Complications

Immediate complications are aspiration of blood (in this case withdraw and resite the needle) and aspiration of CSF.

Beware of intrathecal injections.

N.B.
Electrical stimulation in order to locate the nerve root is not essential, as, owing to the small space, the injected solution will bathe the nerve root. Nevertheless the technique is of value, in as much as it confirms the site of the patient's pain. In this case use a Sluijter–Mehta or pole needle and, with the radiofrequency generator in 'stimulation' mode, try to elicit pain/strong paraesthesiae in the area of the patient's pain at 1 V or less and at a frequency of 100 Hz.

3 Thoracic Nerve Root Injection

Indication

Pain along a thoracic nerve root. The injection may be either diagnostic, perhaps as an indicator of the possible benefits of dorsal root ganglion lesioning, or therapeutic, when a mixture of Bupivacaine and steroid may be injected.

Clinical Picture

The patient presents with pain along a single thoracic nerve. The pain may have all or some of the characteristics of a neuralgia, i.e. burning pain with a shooting component, hypersensitivity, areas of patchy sensation, allodynia, paraesthesiae or dysaesthesiae.

Technique

Sedation

Sedation is not usually necessary.

Position

The patient lies prone on a radiolucent table. With upper thoracic work, it is advisable to arrange the patient's arms so that they hang over the top edge of the table; this manoeuvre lifts the scapulae out of the way (Figure 1).

Landmarks

With C-arm image intensifier screening in the anteroposterior plane, the level of the intervertebral foramen in question is located; the adjacent head of rib is then identified and a skin mark is made in the intercostal space just below the caudal edge of this rib – which forms the rostral boundary of the intercostal space. The mark should be about 5 cm from the midline.

Clean and drape the area.

Procedure

- Infiltrate the entry point with 0.5% Lignocaine with Adrenaline; then direct a 23 gauge 2.5 cm needle towards the intervertebral foramen and infiltrate.
- Insert a 22 gauge 6.25 cm (3½ in) short-bevelled needle through the entry point and direct it towards the vertebral lamina overlying the intervertebral foramen (Figure 2). 'Walk' the needle laterally off the lamina and into the foramen. You will feel a definite 'give' as the needle enters the foramen; advance the needle into the foramen no more medial than an imaginary line joining the facet joints in the anteroposterior view obtained by the image intensifier.
- Turn the image intensifier so as to obtain a lateral view; it is not easy to obtain a true lateral view in the thoracic region, as the ribs get in the way at all levels and the scapulae get in the way in the upper thoracic region. Angle the image intensifier tube so as to get the best lateral (or oblique) view possible. The needle should be lying in the posterior half of the foramen; if necessary, pull the needle back into the posterior half of the foramen (Figure 3). When you are satisfied with the position of your needle, withdraw the stylet and inject.
- If the injection is a *diagnostic* one, inject 2 ml of either 2% Lignocaine or 0.5% Bupivacaine, depending on whether you want to make an immediate or a delayed assessment.
- If the injection is a *therapeutic* one, inject 2 ml of 0.5% Bupivacaine plus your steroid preparation.

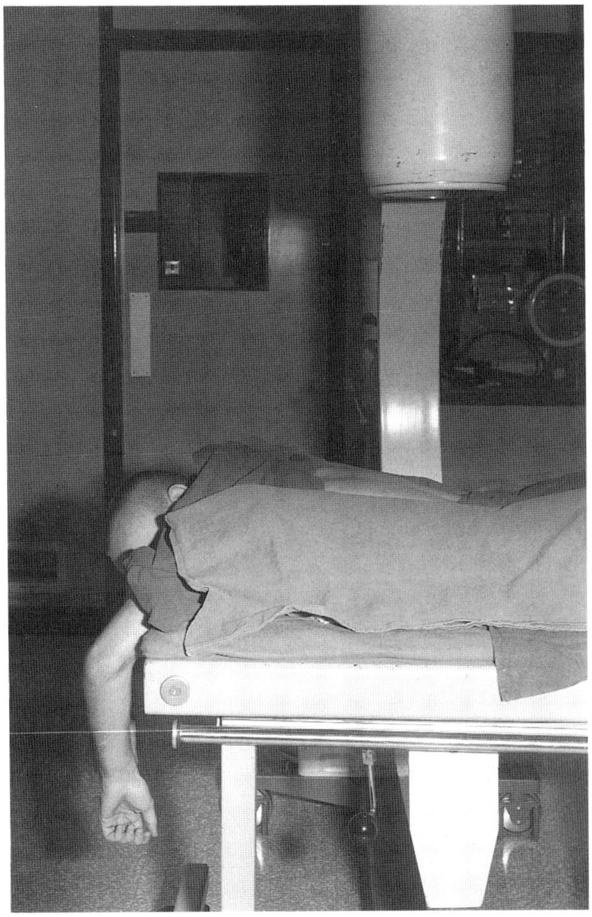

Figure 1. Position of patient for thoracic nerve root injection.

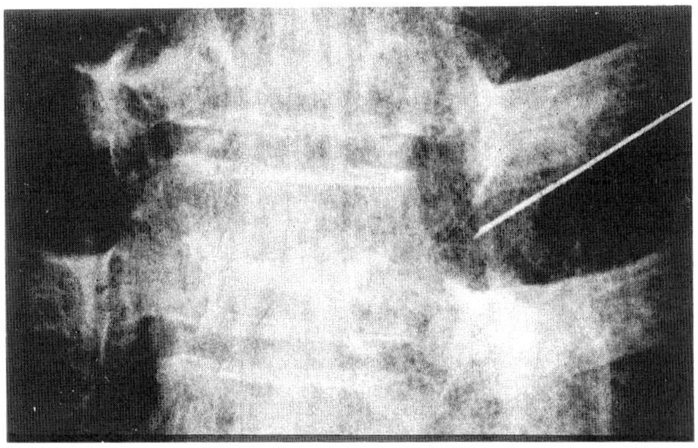

Figure 2. Needle tip approaching thoracic intervertebral foramen.

After-care

It is prudent to keep the patient in hospital for at least 1 h after the lesioning; be on the look-out for the sudden onset of dyspnoea and/or chest pain, which may indicate a pneumothorax.

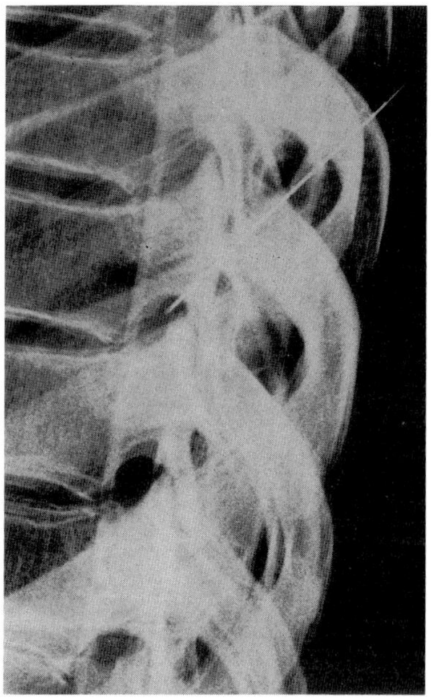

Figure 3. Needle tip in thoracic intervertebral foramen-lateral view.

Complications

Possible immediate complications are:

Aspiration of blood (in this case withdraw and resite the needle).

Aspiration of CSF (less likely to happen if your needle is never advanced past the imaginary line linking the facet joints referred to above).

Pneumothorax (always a possibility the risk of which can be minimised by meticulous technique).

N.B.
Electrical stimulation in order to locate the nerve root is not essential, as, owing to the small space, the injected solution will bathe the nerve root. Nevertheless the technique is of value in as much as it confirms the site of the patient's pain. In this case use a Sluijter–Mehta or pole needle and, with the radiofrequency generator in 'stimulation mode', try to elicit pain/strong paraesthesiae in the area of the patient's pain at 1 V or less and at a frequency of 100 Hz.

4 Lumbar Nerve Root Injection

Indication

- Pain along a lumbar nerve root, especially pain associated with nerve root damage due to compression which has been relieved by laminectomy.
- Pain from post-laminectomy nerve root scarring.

The injection may be either diagnostic, perhaps as an indicator of the possible benefits of dorsal root ganglion lesioning, or therapeutic, when a mixture of Bupivacaine and steroid may be injected.

Clinical Picture

The patient presents with pain along a single nerve root; more often than not, the patient will have had previous back surgery and the surgeons are not keen to re-explore the back.

Technique

Sedation

Sedation is not usually necessary.

Position

The patient lies prone on a radiolucent table with a pillow under the abdomen, in order to straighten the spine (Figure 1).

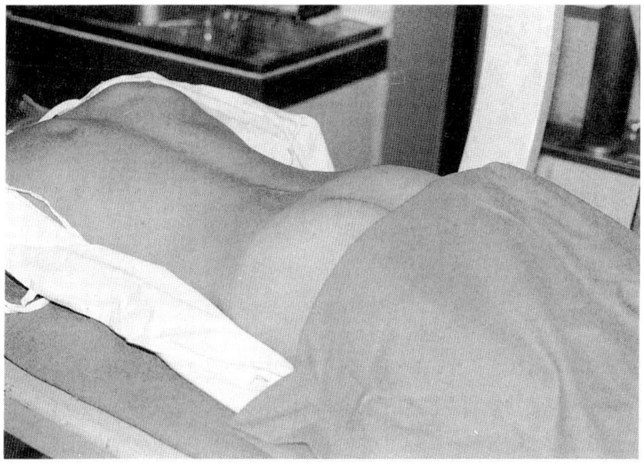

Figure 1. Position of patient for lumbar nerve root injection.

Landmarks

With the C-arm image intensifier positioned for screening in the anteroposterior plane, the level of the intervertebral foramen in question is located. The adjacent transverse process is identified and a skin mark is made below the tip of the process about 2 cm caudad to the intervertebral foramen (Figure 2). This mark is usually about 5 cm from the midline.

Clean and drape the area.

Procedure

- Infiltrate the entry point with 0.5% Lignocaine with Adrenaline; then direct a 23 gauge 2.5 cm needle towards the foramen and infiltrate with local anaesthetic.
- Insert a 20 gauge 6.5 cm (3½ in) spinal needle through the entry point and direct it towards the vertebral lamina overlying the intervertebral foramen; 'walk' the needle off the lamina and into the foramen. You will feel a definite 'give' as the needle enters the foramen. Advance the needle into the foramen no more medial than an imaginary line joining the facet joints in the anteroposterior view obtained on the image intensifier (Figure 3). If you experience difficulty in inserting the needle into the foramen, try a more lateral entry point.
- Rotate the image intensifier so as to obtain a lateral view of the spine; the needle should be lying in the posterior half of the intervertebral foramen; if necessary, pull the needle back into the posterior half of the foramen (Figure 4).

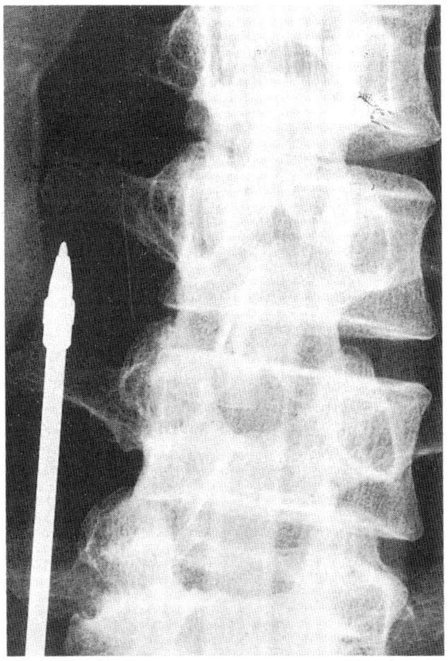

Figure 2. Marker over skin entry point for lumbar nerve root injection.

- If the injection is a *diagnostic* one, inject 2 ml of either 2% Lignocaine or 0.5% Bupivacaine, depending on whether you want to obtain an immediate short-term or a delayed, more prolonged effect.
- If the injection is a *therapeutic* one, inject 1 ml of 0.5% Bupivacaine plus your steroid preparation.

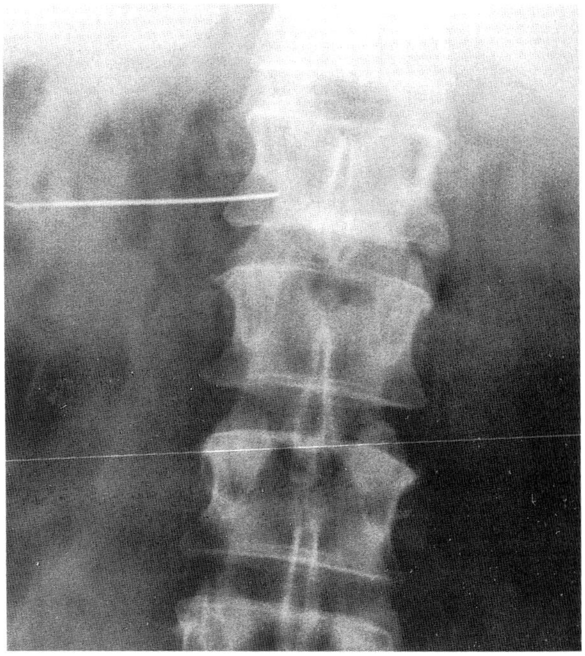

Figure 3. Needle tip in lumbar intervertebral foramen-postero/anterior view.

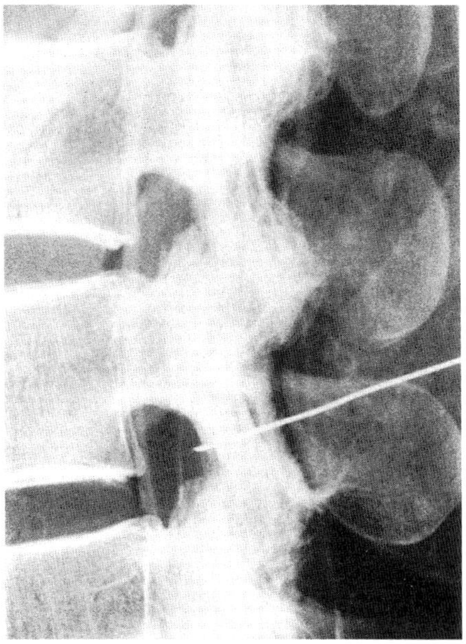

Figure 4. Needle tip in lumbar intervertebral foramen-lateral view.

After-care

It is prudent to keep the patient in hospital for at least 1 h after the injection.

Complications

Possible immediate complications are:

Aspiration of blood (withdraw and resite the needle).

Aspiration of CSF (this is less likely to happen if your needle is never advanced past the imaginary line linking the facet joints referred to above).

> **N.B.**
> Electrical stimulation in order to locate the nerve root is not essential, as owing to the small space, the injected solution will bathe the nerve root. Nevertheless the technique is of value, in as much as it confirms the site of the patient's pain. In this case use a Sluijter–Mehta or pole needle and, with the radiofrequency generator in 'stimulation mode', try to elicit pain/strong paraesthesiae in the area of the patient's pain at 1 V or less and at a frequency of 100 Hz.

5 Sacral Nerve Root Injection

Indication

- Pain along a sacral nerve root, especially pain associated with S1 nerve root damage due to compression which has been relieved by laminectomy.
- Pain from post-laminectomy nerve root scarring.

The injection may be either diagnostic, perhaps as an indicator of the possible benefits of sacral nerve rhizolysis (see page 20), or therapeutic, when a mixture of Bupivacaine and steroid may be indicated.

- Pain of malignant origin along the pathways of the lower sacral nerve roots; in this case it is usually done as a diagnostic test prior to rhizolysis.

Clinical Picture

- S1 pain: the patient presents with pain along the distribution of the first sacral nerve. More often than not, the patient has had previous back surgery and the surgeons are not very keen to re-explore the back! The pain may have all or some of the characteristics of a neuralgia, i.e. burning pain with a shooting component, hypersensitivity, areas of patchy sensation, allodynia, paraesthesiae and dysaesthesiae.
- S2–S4 pain: pain along the lower sacral nerve roots, usually neuralgic in character and exhibiting the same features as described above.

Technique

Sedation

Sedation is not necessary.

Position

The patient lies prone on a radiolucent table with a pillow under the abdomen in order to straighten the spine.

Landmarks

With the C-arm image intensifier positioned for screening in the anteroposterior axis (Figure 1a), the level of the L5/S1 intervertebral foramen and the sacral foramina is identified; the image intensifier should now be positioned so that it is at exactly right angles to the sacrum (Figure 1b).

Clean and drape the area.

Procedure

There are two possible entry points for S1 root injection, viz. directly through the S1 foramen; or through the L5/S1 foramen.

Through the S1 Foramen

Mark the position of the first sacral foramen on the skin.

- Infiltrate the entry point with 0.5% Lignocaine with Adrenaline; insert a 20/22 Gauge 8.25 cm (3½ in) spinal needle into the foramen (Figure 2).
- Rotate the image intensifier so as to obtain a lateral view; the needle should be lying in the sacral canal (Figure 3).
- If the injection is a *diagnostic* one, inject 2 ml of either 2% Lignocaine or 0.5% Bupivacaine, depending on whether you want an immediate short-term or a delayed, more prolonged effect.

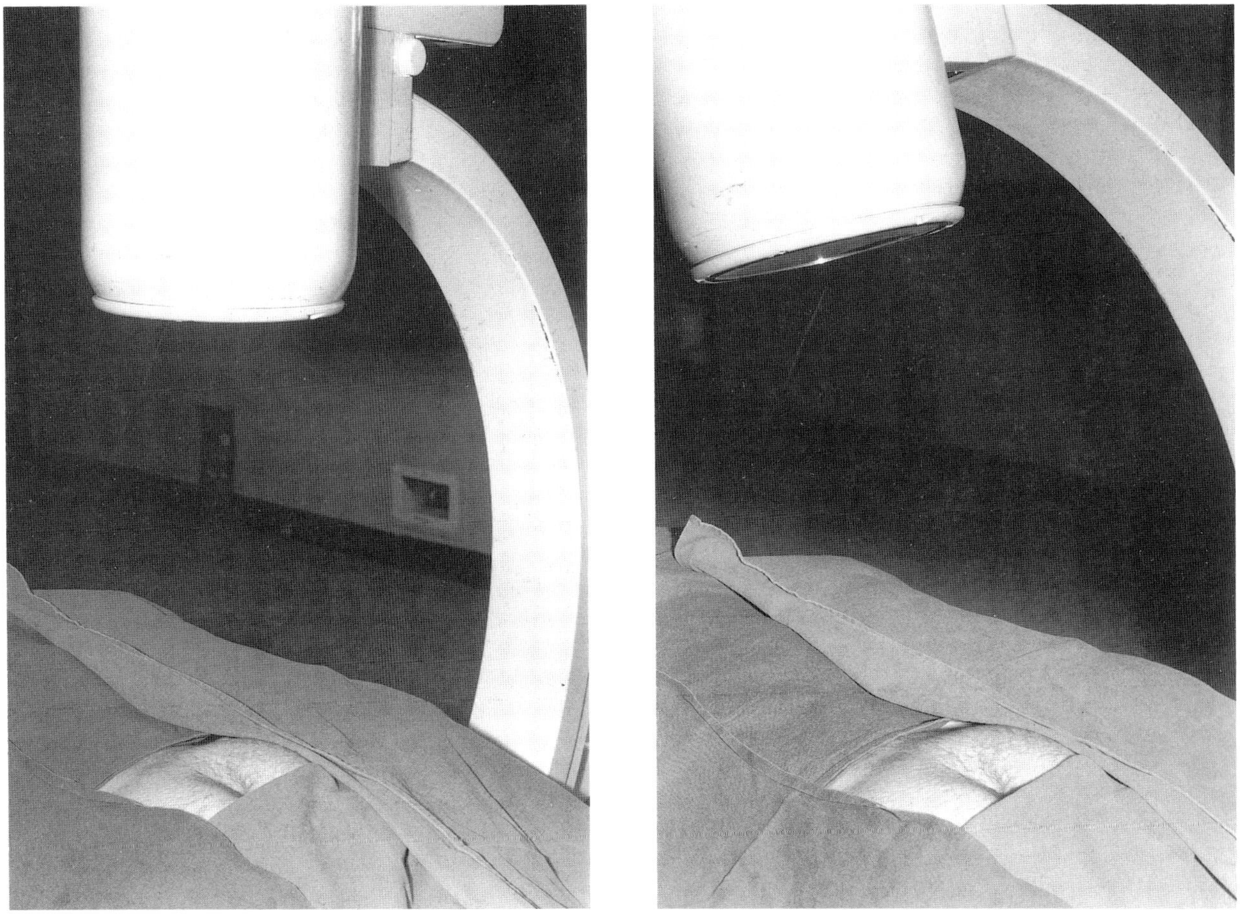

Figures 1a and 1b. Positioning the image intensifier to visualise the sacral foramina.

- If the injection is a therapeutic one, inject 2 ml of 0.5% Bupivacaine plus your steroid preparation.

Through the L5/S1 Foramen

Insert your needle through the L5/S1 intervertebral foramen, using the technique described on page 3, but angle the needle caudally and advance it so as to make contact with the S1 root (Figure 4). Proceed as above.

The S2–S4 Nerve Roots

If you wish to inject these roots, insert your spinal needle through the sacral hiatus and direct it laterally, although, in truth, you may as well carry out a caudal epidural injection.

After-care

It is prudent to keep in hospital for at least 1 h after the injection.

Complications

(Immediate) aspiration of blood: withdraw and resite the needle.

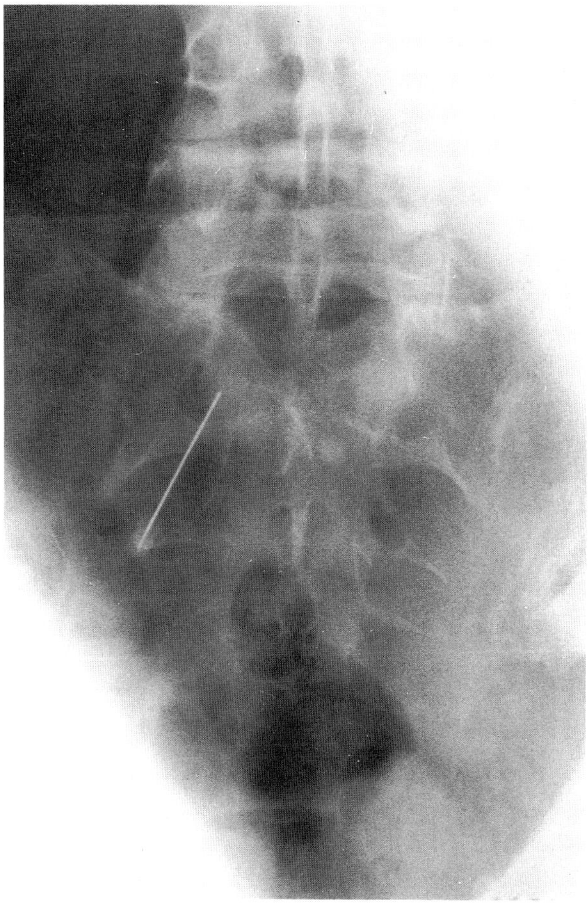

Figure 2. Needle in S1 foramen-postero/anterior view.

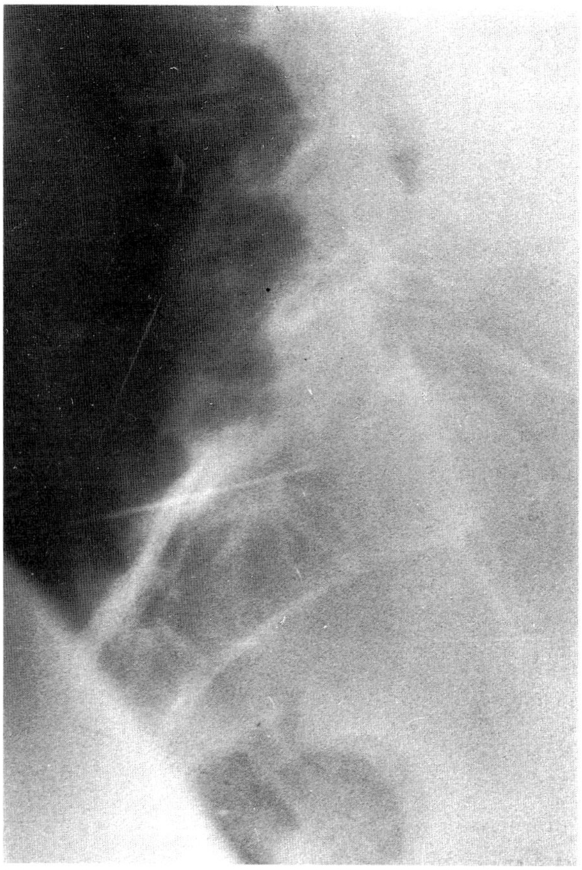

Figure 3. Needle in S1 foramen-lateral view.

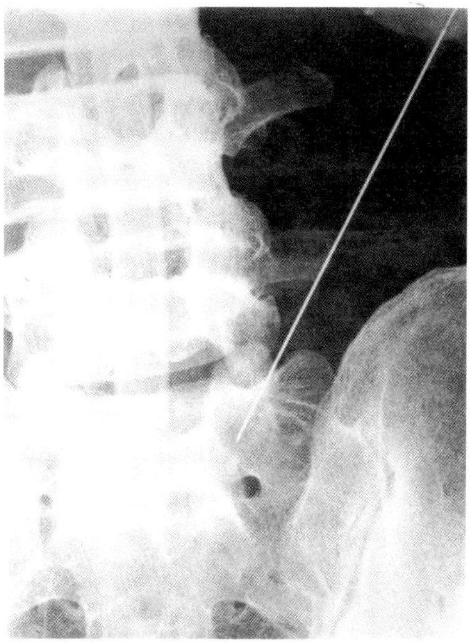

Figure 4. Needle in introduced onto S1 nerve root via L5/S1 intervertebral foramen.

N.B.
Electrical stimulation in order to locate the nerve root is not essential, as, owing to the small space, the injected solution will bathe the nerve root. Nevertheless the technique is of value, in as much as it confirms the site of the patient's pain. In this case use a Sluijter–Mehta or pole needle and, with the radiofrequency generator in 'stimulation mode', try to elicit pain/strong paraesthesiae in the area of the patient's pain at 1 V or less and at a frequency of 100 Hz.

6 PSOAS Compartment Block

Indication

Hip pain, especially in patients waiting for hip replacement surgery.

Technique (2)

Sedation
Sedation is usually necessary.

Position
The patient lies in the lateral position, the painful side uppermost; thighs should be flexed on hip.

Landmarks
Identify the spine of L4 by a line passing through both iliac crests; mark a point 3 cm caudad to the spine and then a second point 5 cm lateral to the first point (Figure 1). This is your entry point.

Clean and drape the area.

Procedure
- Infiltrate the entry point with 0.5% Lignocaine with Adrenaline, and then insert a 15 cm 20 gauge needle perpendicularly through the skin until it strikes the transverse process of L5 (Figure 2).
- 'Walk' the needle cephalad until it just passes above the transverse process. Attach a 20 ml syringe containing air to the hub of the needle and then advance it for 1–2 cm until it enters the quadratus lumborum muscle; this point is easily identified, as a 'pneumatic bounce' is obtained by gentle pressure on the plunger when the muscle or its fascia has been entered.

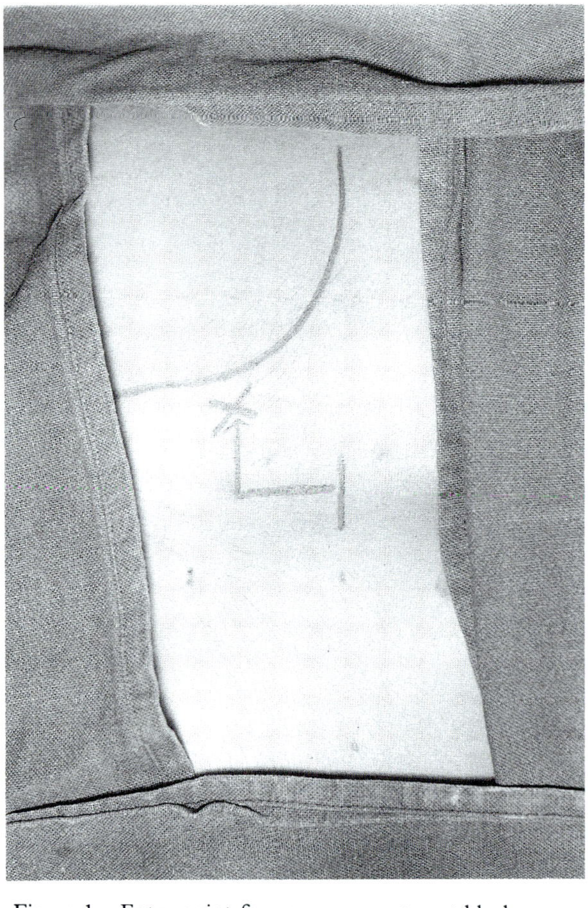

Figure 1. Entry point for psoas compartment block.

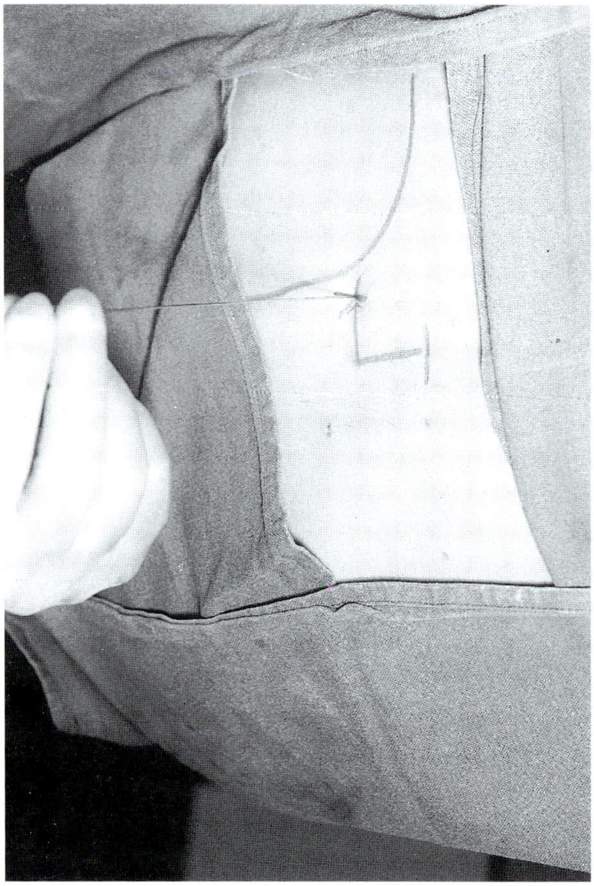

Figure 2. Insertion of needle in psoas compartment block.

- Advance the needle slowly until the 'pneumatic bounce' disappears; the tip of your needle is now in the psoas compartment, which contains the lumbar plexus.
- Aspirate to ensure that you have not entered a blood vessel, and inject 20 ml of air to dilate the compartment; inject 30 ml of 0.375% Bupivacaine plus your steroid preparation.

After-care

Leave the patient on his or her side for about 5 min.

(1) Gallagher, J. and Hamann, W. (1989) Chronic neuropathic pain: aminoglycosides, peripheral somatosensory mechanisms and painful disorders. In *Recent Advances in Anaesthesia and Analgesia*, No. 16 Churchill Livingstone, 191–205.

(2) The authors use the technique described by Chayen D., Nathan H. and Chayen, M. (1976) in *Anesthesiology*, **45**, No. 1, 95–99.

9

Joint Injections

1 Cervical Facet Joint Injection

Indication

Cervical facet joint pain.
(1) Diagnostic; (2) therapeutic.

Clinical Picture

The patient presents with a stiff and painful neck, sometimes accompanied by a frank torticollis. The pain commonly radiates into the back of the head and sometimes over the shoulder and down the arm in a non-radicular fashion. Pain is aggravated by movement and is frequently positional. This often causes difficulty in finding a comfortable position in which to sleep. Pain is relieved by immobilisation.

On examination cervical flexion is normally full and free. Extension, lateral rotation and lateral flexion are limited and painful. Muscle spasm is often present. Neurological examination is usually normal.

Technique

Sedation
Sedation is not usually necessary.

Position
The patient lies supine with the head on a radiolucent rest which affords full access to the neck; a C-arm image intensifier is positioned as shown in Figure 1.

Landmarks
Identify the mastoid process and mark its apex; draw a line from here down along the posterior border of the sternomastoid. One finger's breadth below the apex of the mastoid process corresponds approximately with the entry point for the C2/C3 facet joint.

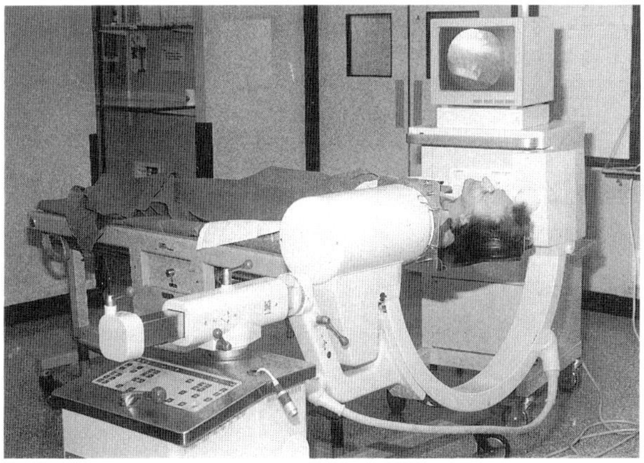

Figure 1. Initial position of image intensifier in cervical facet joint injection.

View the cervical spine with the image intensifier positioned as shown in Figure 1; note that the intensifier end of the C-arm is situated to face the side being lesioned, while the X-ray tube part is situated on the opposite side. You now have a lateral view of the spine (Figure 2). Rotate the C-arm (Figure 3) so as to obtain an oblique view of the spine; in this view, the intervertebral foramina and their associated facet joints are clearly seen (Figure 4).

Clean and drape area.

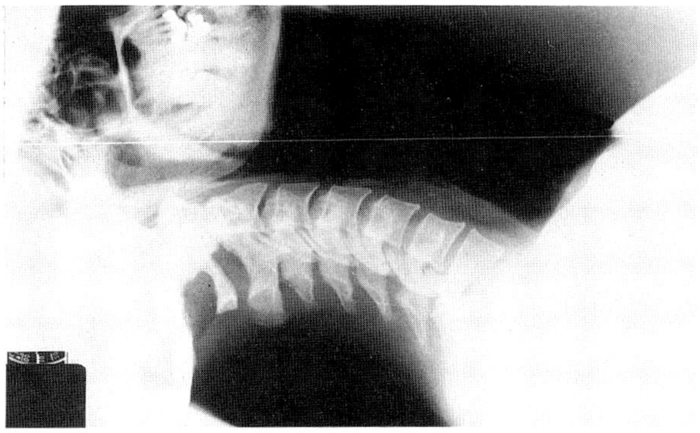

Figure 2. Lateral view of cervical spine with image intensifier.

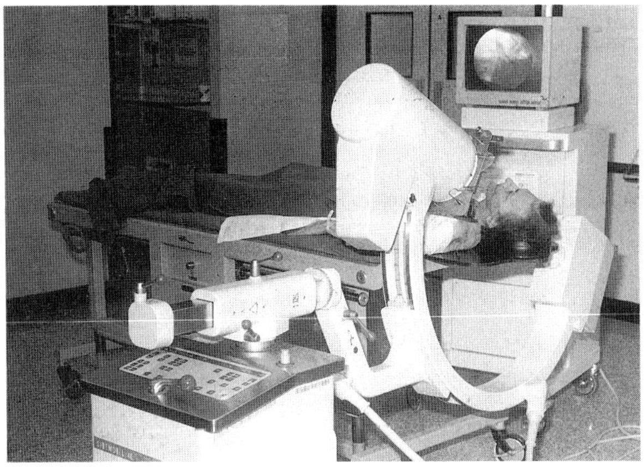

Figure 3. Final position of image intensifier in cervical facet joint injection.

Procedure

- Infiltrate the entry point with 0.5% Lignocaine with Adrenaline (Figure 4).
- Insert a 21 gauge 5 cm (2 in) needle, aiming for the facet joint in question; the needle should be placed in the facet joint about 1–2 mm dorsal to the posterior edge of the intervertebral foramen (Figure 5).
- If the block is diagnostic, inject 2 ml of 0.5% Bupivacaine; the longer duration of action gives it an advantage over Lignocaine in cervical pain, as patients have a couple of hours in which to assess whether or not their neck pain has improved. If the block is therapeutic, inject 1 ml of 0.5% Bupivacaine plus 1 ml of your steroid preparation. When finished, withdraw the needle and move on to the facet below. The entry point for your needle should be about one finger's breadth caudad to the point above.

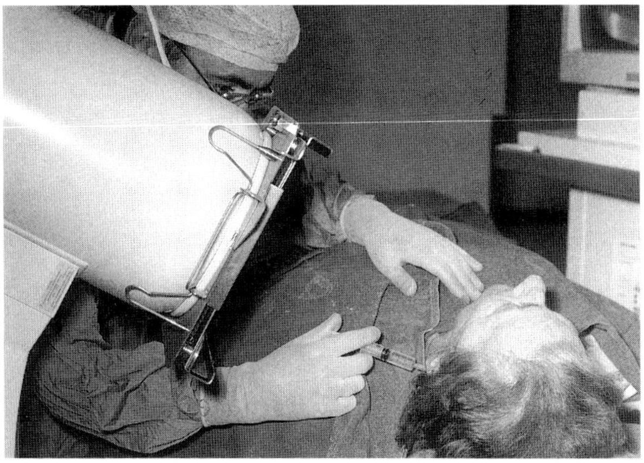

Figure 4. Infiltrating the entry point.

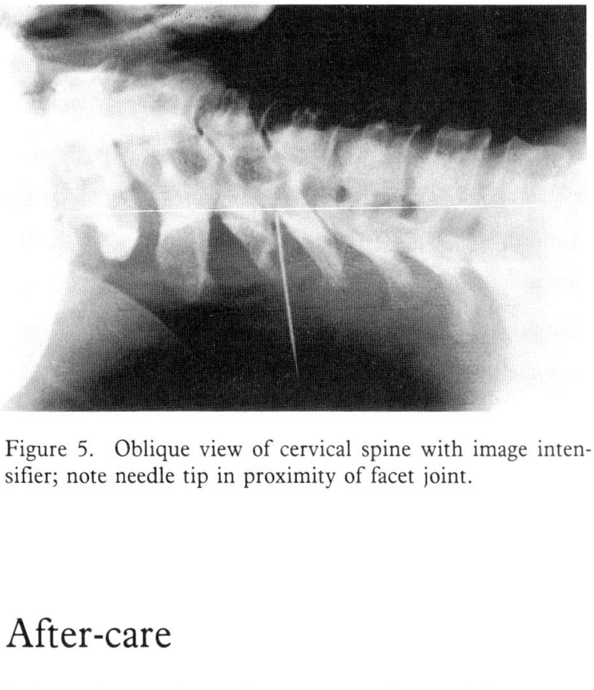

Figure 5. Oblique view of cervical spine with image intensifier; note needle tip in proximity of facet joint.

After-care

It is prudent to keep the patient in hospital for at least 1 h after the injection; observe in case you have accidentally injected Lignocaine/Bupivacaine into the CSF!

Complications

Possible immediate complications are aspiration of blood (withdraw and resite the needle); aspiration of CSF (this should not happen if your needle is always in contact with bone).

2 Thoracic Facet Joint Injection

Indication

Thoracic facet joint pain.

Clinical Picture

The patient presents with diffuse pain over the thoracic spine; the pain is often unilateral, radiating laterally over the thorax, and there may be muscle spasm present. The patient finds it uncomfortable to lie back in bed or to sit in a chair. Minimal activity aggravates the pain.

On examination there is often some protective muscle spasm. As a rule, no clinical signs are present, except for tenderness in the paravertebral space overlying the affected facets.

Technique

Sedation

Sedation is not usually necessary.

Position

The patient lies prone on a radiolucent table, with the C-arm image intensifier positioned as in Figure 1.

Landmarks

It is impossible to inject material into the thoracic facet joints: you have to make do with a periarticular injection. With the image intensifier screening in the antero-posterior plane, locate the facet joint in question and mark the overlying skin with a pen.

Procedure

After cleaning and draping the area and infiltrating with 2% Lignocaine, direct a 22 gauge 8.75 cm (3½ in) needle to the area where the head of the rib meets the thoracic spine at the facet level in question (Figure 2).

If the block is *diagnostic*, inject 2 ml of 0.5% Bupivacaine or 2 ml of 2% Lignocaine, depending on whether you want an immediate short-term or a delayed effect. If the block is *therapeutic*, inject 1 ml of 0.5% Bupivacaine plus 1 ml of your steroid preparation.

After-care

It is prudent to keep the patient in hospital for 1 h after the injection.

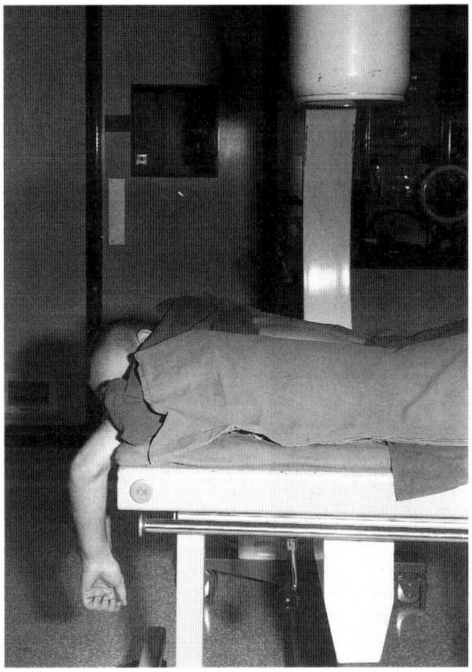

Figure 1. Position of patient for thoracic facet joint injection.

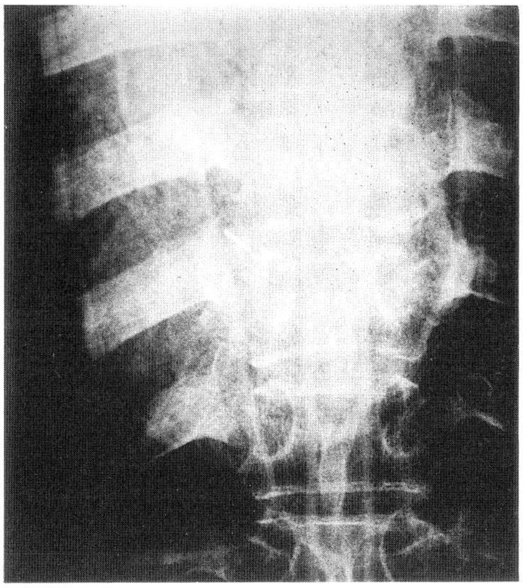

Figure 2. Thoracic facet joint injection.

Complications

Possible immediate complications are aspiration of blood (withdraw and resite the needle); aspiration of CSF (this should not happen if your needle is always aimed at the target zone specified above); pneumothorax (totally avoidable if you keep your needle in contact with bone).

3 Lumbar Facet Joint Injection

Indication

Lumbar facet joint pain.
(1) Diagnostic; (2) therapeutic.

Clinical Picture

The patient presents with a stiff and painful back. The back pain is often accompanied by pain in the buttocks and thighs, but does not normally extend below the knees, although it can do so. The pain is exacerbated by prolonged sitting and standing, and the patient normally finds it very difficult to get a good night's sleep because of the pain. Coughing and sneezing do not usually exacerbate the pain. The patient prefers to be mobile rather than still.

On examination there is often much protective muscle spasm. Forward flexion of the lumbar spine is usually full and free, but it may sometimes be restricted. Extension is invariably limited by sharp pain. Lateral flexion and lateral rotation may also be limited by pain.

The facet joints are very tender to palpation.

Straight leg raising is usually normal but may be accompanied by pain in the lower back. There are no root signs in pure facet joint pain. Neurological examination is normal.

Lumbar X-rays and CT scans are usually normal or perhaps show varying degrees of facet joint degeneration, but there is usually no correlation between the radiological appearance and the presence or intensity of the pain.

If the patient has had a lumbar fusion, it is not at all uncommon for him to develop facet joint pain above the fusion level.

Technique

Sedation

Sedation is not usually necessary.

Position

The patient lies prone initially with the image intensifier screening in the anteroposterior plane.

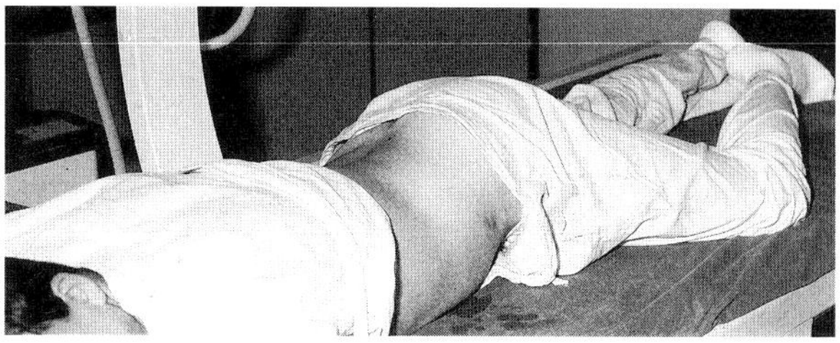

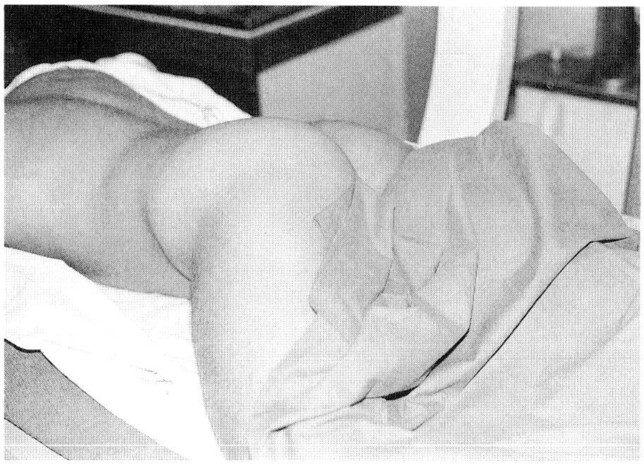

Figures 1a and 1b. Positioning the patient to visualise the left lumbar facet joints.

Two Techniques

1. The X-ray beam is kept vertical and the patient is gently moved into an oblique position so that the facet joints appear prominently into view (Figure 1a, b). The obliquity is achieved by flexing the patient's thigh and bending the knee (on the affected side) so that the patient adopts the 'frog's leg' position. The patient is rotated so that the affected side is uppermost. The joint which may sometimes be difficult to visualise is the L5/S1 facet, as it is often hidden by the iliac crest; the patient must be rotated until the joint is seen just medial to the inferior part of the crest (Figure 2). This technique is the only suitable technique if your equipment has a fixed X-ray beam, e.g. a screening unit, not a C-arm image intensifier.

2. The patient is also initially moved into an oblique position as above, but 'fine tuning' is achieved by moving the X-ray beam off the central anteroposterior axis into an oblique position.

Entry Points

Mark the skin over the facet joint you wish to inject. The area must be cleaned and draped.

Procedure

- Infiltrate the entry points with 0.5% Lignocaine with Adrenaline.

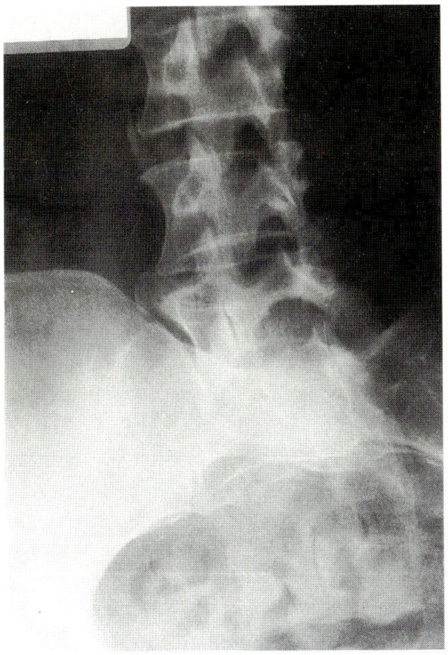

Figure 2. The left lower lumbar facet joints with patient positioned as shown in Figs. 1a & 1b and image intensifier screening in antero/posterior plane.

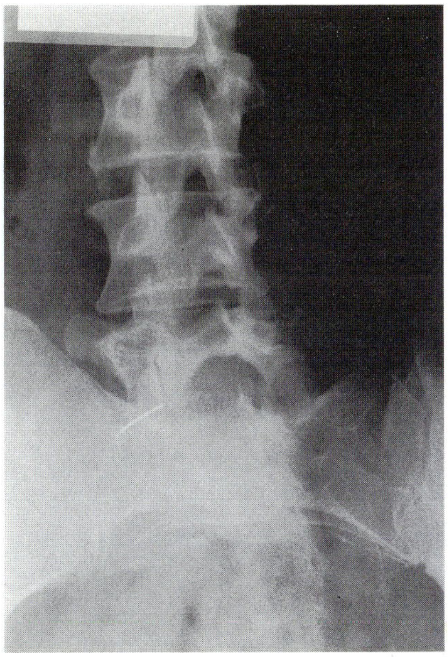

Figure 3. Needle tip in L5/S1 facet joint.

- Insert a 20/22 gauge 8.5 cm (3½ in) spinal needle (the gauge will depend upon the patient's build) along the axis of your X-ray tube into the joint, checking your position radiologically as you go deeper and deeper. Very often you will feel a 'give' as you enter the joint (Figure 3).

Injection: diagnostic, either 2 ml of 0.5% Bupivacaine or 2 ml of 2% Lignocaine; therapeutic, 1 ml of 0.5% Bupivacaine together with 1 ml of your steroid preparation.

It is very important that you aspirate before injecting, as the area is very vascular indeed.

After-care

None.

Complications

Possible immediate complications are aspiration of blood (withdraw and resite the needle); aspiration of CSF (this should not happen if you use frequent screening to check the position of your needle).

4 Sacro Iliac Joint Injection

Indication

Sacroiliac joint pain.

Clinical Picture

The patient presents with a stiff and painful back. The back pain is often accompanied by pain in the buttocks and thighs, but does not normally extend below the knees, although it can do so. The pain is exacerbated by prolonged sitting and standing, and the patient normally finds it very difficult to get a good night's sleep. Coughing and sneezing do not usually exacerbate the pain. The patient prefers to be mobile rather than still; however, movement may also exacerbate his or her pain. This condition is sometimes difficult to distinguish from L5/S1 facet joint pain, but you can usually elicit tenderness over the sacroiliac joints.

On examination flexion of the lumbar spine is usually full and free but it may sometimes be restricted. Extension is invariably limited by sharp pain. The sacroiliac joints are very tender to palpation; the straight leg raising test is usually normal but may be accompanied by pain in the lower back. There are no root signs and neurological examination is normal. Lumbar X-rays may be of value, as are bone scans (sacro-iliitis).

Technique

Sedation

Sedation is not necessary.

Position

Initially, the patient lies prone on a radiolucent table with the image intensifier positioned for screening in the anteroposterior plane; the operator can familiarise himself with the general position of the joints.

Two Techniques

1. The X-ray beam is kept vertical and the patient is gently moved into an oblique position so that the sacroiliac joint appears prominently into view. The obliquity is achieved by flexing the patient's thigh and bending the knee (on the side *opposite* to the side you want to inject) so that the patient adopts the 'frog's leg' position. The patient is moved so that the affected side is lowermost; by gently adjusting the patient's position, the optimum position for injection (i.e. maximum width of the 'joint line' will become visible (Figure 1). This technique is the only suitable technique if your equipment has a fixed X-ray beam, e.g. a screening unit, not a C-arm image intensifier.

2. The patient is kept prone and the X-ray beam is moved so that it is exactly at right angles to the sacrum; this should give you a good view of the sacroiliac joint. It is still, however, rather helpful to obtain the oblique view described above.

Entry Points

Mark the skin over the joint you wish to inject; you may either carry out a single injection into the joint or deposit your solution at several points along either side of the joint space. The area must be cleaned and draped.

Procedure

- Infiltrate the entry point with 0.5% Lignocaine with Adrenaline.

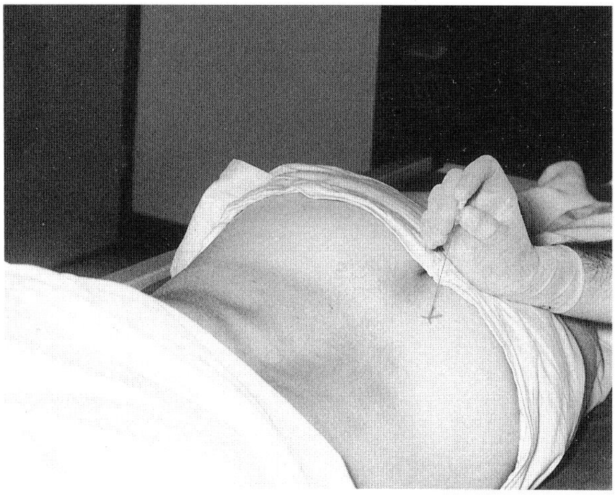

Figure 1. Position of patient for left sacroiliac joint injection.

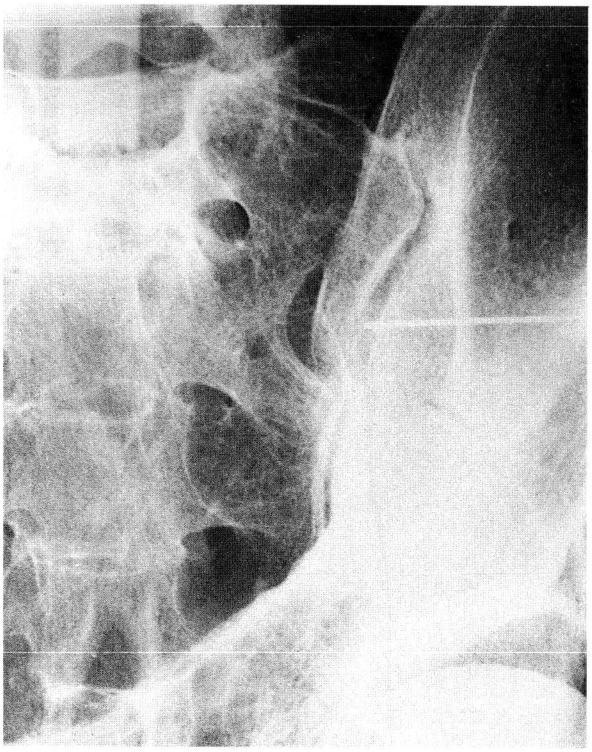

Figure 2. Needle tip in right sacroiliac joint.

- Insert a 20/22 gauge 8.5 cm (3½ in) spinal needle into the joint, checking your position radiologically as you go deeper and deeper. Very often you will feel a 'give' as you enter the joint (Figures 1, 2).

Injection: diagnostic, either 2 ml of 0.5% Bupivacaine or 2 ml of 2% Lignocaine, depending on whether you want an immediate short-term or a more prolonged effect; therapeutic, 5 ml of 0.5% Bupivacaine with 50 mg of Hydrocortisone Acetate is suitable for a single joint injection (if you prefer to inject at several points, then 80 mg of Methylprednisolone made up to 10 ml with 0.5% Bupivacaine is a good alternative).

It is very important that you aspirate before injection, as the area is very vascular.

After-care

None.

Complications

Possible immediate complications are aspiration of blood (withdraw and resite the needle); aspiration of CSF (this should not happen if you repeatedly screen to check the position of your needle.